Lecture Notes in Control and Information Sciences

Patricia Mellodge, Pushkin Kachroo

Model Abstraction in Dynamical Systems: Application to Mobile Robot Control

Authors

Dr. Patricia Mellodge

Department of Electrical and Computer Engineering
University of Hartford
200 Bloomfield Avenue
West Hartford, CT 06117
USA
E-Mail: mellodge@hartford.edu

Dr. Pushkin Kachroo

University Transportation Center
Howard R. Hughes College of Engineering
4505 Maryland Parkway
Las Vegas, NV 89154-4007
USA
&
Department of Electrical and Computer Engineering
University of Nevada, Las Vegas
4505 S. Maryland Pkwy
Las Vegas, NV 89154-4026
USA
E-Mail: pushkin@vt.edu

ISBN 978-3-540-70792-9 e-ISBN 978-3-540-70799-8

DOI 10.1007/978-3-540-70799-8

Lecture Notes in Control and Information Sciences ISSN 0170-8643

Library of Congress Control Number: 2008930993

Typeset & Cover Design: Scientific Publishing Services Pvt. Ltd., Chennai, India.

Printed in acid-free paper

5 4 3 2 1 0

springer.com

Contents

List of Figures

List of Tables

Acknowledgments

Portions of Chapter 3 and Chapter 4 appear in the McGraw-Hill book *Mobile Robotic Car Design* by Kachroo and Mellodge, copyright 2005.

1 Introduction

The subject of this book is model abstraction of dynamical systems. The primary goal of the work embodied in this book is to design a controller for the mobile robotic car using abstraction. Abstraction provides a means to represent the dynamics of a system using a simpler model while retaining important characteristics of the original system. A second goal of this work is to study the propagation of uncertain initial conditions in the framework of abstraction. The summation of this work is presented in this book. It includes the following:

- An overview of the history and current research in mobile robotic control design.
- A mathematical review that provides the tools used in this research area.
- The development of the robotic car model and both controllers used in the new control design.
- A review of abstraction and an extension of these ideas into new system relationship characterizations called *traceability* and *ϵ-traceability.*
- A framework for designing controllers based on abstraction.
- An open-loop control design with simulation results.
- An investigation of system abstraction with uncertain initial conditions.

1.1 Motivation

One of the major motivations for studying mobile robotic cars is to automate the driving task. In 2006, there were 5,973,000 police-reported motor vehicle traffic crashes, in which 42,642 involved fatalities and 2,575,000 involved injury [42]. There is evidence to show that many automobile crashes are the result of driver distraction or fatigue. According to a study conducted by the Virginia Tech Transportation Institute [41], driving while drowsy was a factor in 22% to 24% of crashes and near-crashes and distracted driving was a factor in more than 22% of crashes and near-crashes. Alleviating humans from having to control all aspects of driving may reduce fatalities and injuries on the roadways.

Automobile manufacturers have developed and are continuing to develop systems for cars that alleviate the driver's need to monitor and control all aspects

P. Mellodge & P. Kachroo: Model Abstraction in Dynamical Systems, LNCIS 379, pp. 1–5, 2008.
springerlink.com

of the vehicle. Such systems include antilock braking systems, traction control, and cruise control. As each of these features was introduced, the burden of controlling the vehicle was placed less on the human driver and more on the car's control system. A completely autonomous vehicle is one in which a computer performs all the tasks that the human driver would normally perform. Ultimately, this would mean getting in a car, entering the destination into a computer, and enabling the system. From there, the car would take over and drive to the destination with no human input. The car would be able to sense its environment and make steering and speed changes as necessary.

Another reason to automate cars is to alleviate congestion on the highways. A method called "platooning" allows cars to drive at highway speeds while only a few feet apart. Since the electronics on the car can respond faster than a human, cars would be able to drive much closer together. This allows much more efficient use of the existing highways in a safe manner.

The control of mobile robotic cars also has applications in areas where humans can or should not go. For example, robots could be used to navigate other planets. It is safer and less expensive to send a relatively small mobile robot to the harsh environment of Mars than develop transport and living quarters for several astronauts. Similarly, mobile robots could be used to enter burning buildings to locate someone or something, navigate battlefields to search for mines, or seek out and deactivate bombs.

The applications can be expanded if one views the system not as a robotic car, but rather as a more general system. If the unimportant details of the car are abstracted out and only the essential details are encompassed in the model, then the robotic car is a nonholonomic system. Other examples of nonholonomic systems are underwater vehicles, walking robots, robotic hand manipulators, and airplanes. Since all of these systems share certain characteristics, the results of studying nonholonomic systems can be applied to all of these different systems.

1.2 Previous Research

1.2.1 Modeling

Before one can control a system, a model is needed. Robotics researchers have taken two different approaches to modeling mobile robots: kinematic and dynamic. In a kinematic model, only the movement of the vehicle is considered and it is derived using the nonholonomic constraints inherent in the vehicle. These constraints are derived in detail in Chapter 3. As will be shown, the kinematic model is fairly easy to derive and is simple in form, owing to its popularity. However, this model fails to account for acceleration and can differ greatly from the actual movement when the vehicle's handling limits are approached. Because of its simplicity, many researchers use the kinematic model and place the emphasis on designing robust controllers. In [3], the relationship between the robot's rigid motion and the wheel's steering and drive rates are derived. The results in [32] address the more general system of two constrained rigid rolling bodies.

On the other hand, dynamic modeling accounts for the properties of the vehicle related to its acceleration, such as mass, center of gravity, etc. This type of modeling is much more true to the actual behavior of the vehicle, but the resulting system representation is much more complicated. See [25] and [23], for example.

The model used throughout this work is the kinematic model. This approach is justified by the decoupled architecture described in Chapter 3 in which the vehicle dynamics are accounted for by a controller operating at a lower level.

1.2.2 Nonholonomic Motion Planning and Control

The goal of motion planning is to find a path for a robot to follow through an environment containing obstacles. See [10] for an overview. A major area of research utilizing kinematic modeling is nonholonomic motion planning. This type of motion planning is an extension of the "Piano Movers' Problem" in which the obstacle positions are known, but the dynamic constraints of the system are ignored [33]. Nonholonomic motion planning accounts for the velocity constraints during the design process.

One of the first important results in nonholonomic motion planning, due to Brockett [9], gives necessary conditions for a system to have smooth stabilizing inputs. Applying this condition to the car, there does not exist a smooth time-invariant input to stabilize the robotic car to a point. Later, Murray and Sastry developed an approach extending Brockett's optimal control results [8] using sinusoids for steering systems in a canonical form called chained form [39].

There are several researchers who have contributed to the study of nonholonomic motion planning and control. Brockett and Sussmann have done much work to establish the controllability of systems using geometric techniques. For example, see [9], [8], [7], [56], [57], and [55]. Later, Bellaiche and Laumond applied these results to robotic cars with n trailers [6][5]. Laumond [29] established the controllability of the n-body system. Samson, De Luca, and Oriolo derived feedback control laws for the car with n trailers. Samson [52] utilized chained form to design a time-varying feedback controller, while Oriolo and De Luca [45] employed feedback linearization to create a stabilizing controller for trajectory tracking and point regulation. In [34], several controllers are presented to perform trajectory tracking, path following, and point-to-point stabilization. An comprehensive overview of nonholonomic robotic motion planning and control is given in [30].

Recently, the concept of abstraction has appeared in the literature as a means of creating control system hierarchies. For example, see [47] and [48]. The hierarchical structure is created by grouping states into equivalence classes. This approach seeks to capture the important characteristics of a system using a simpler (abstracted) system. Of particular importance is controllability. In [47], conditions are given for controllability to propagate between a complex system and its abstraction. This topic is addressed in detail in Chapter 5.

1.2.3 Other Control Approaches

Several researchers have developed controllers for nonholonomic mobile robots using various design techniques. A few of the techniques are Lyapunov-based design, neural networks, wavelet networks, backstepping, and sliding mode control. In [2], the state variables were chosen so that a simple quadratic function served as a Lyapunov function that led to a feedback control law for path following. A neural network controller is given in [13] in which the network learns the dynamics of the robot online. Another online adaptive controller using wavelet networks is proposed in [54] to account for dynamics not modeled in the kinematic model. In [31], a controller is developed using backstepping and LaSalle's invariance principle. A sliding mode controller is presented in [62] to perform trajectory tracking.

1.3 Contributions of This Work

This work has two main contributions:

- The concept of abstraction is used to design a controller for a system based on an abstraction of its model. This concept is applied to the car/unicycle system and it shown that previous characterizations of abstraction do not fully capture certain behaviors that must be understood for control design.
- Abstraction of systems with uncertain initial conditions is studied in detail and it is shown that a commutative relationship exists between the systems. This shows that abstraction can be extended to the stochastic setting.

1.4 Organization of This Work

This book is organized as follows:

- Chapter 2 gives mathematical background necessary for understanding control theory in a geometric context. It begins by describing the ideas such as manifolds, tangent vectors, and vector fields, and important results such as Frobenius' Theorem. It then discusses control system properties, nonholonomic systems, and chained form.
- Chapter 3 describes the kinematic modeling of the car-like robot. Several different models are derived and a controller from [34] is presented. Also, a curvature estimation scheme to be used by the controller is provided. Simulation results are given for both the controller and the curvature estimation.
- Chapter 4 gives another model for the car-like robot with an attached camera. A controller based on this model is presented from [35] and simulation results are given.
- Chapter 5 reviews the concept of consistent abstraction from [47] and introduces the new concepts called traceability and ϵ-traceability. These concepts are applied to the car-like robot by using the unicycle model as an abstraction to show shortcomings in previously defined notions of abstraction. A

comparison between these new concepts and previous characterizations is given.

- Chapter 6 provides a framework for control design using abstraction. The mathematical details are given for general control systems and they are applied to the car/unicycle example.
- Chapter 7 develops an open-loop control scheme based on abstraction. The model for the unicycle is used as a tracking reference for the car-like robot. An optimal control algorithm is developed and simulation results are given.
- Chapter 8 shows that given a system and its abstraction, the evolution of uncertain initial conditions in the original system is matched by the evolution of the uncertainty in the abstracted system. These ideas rely on the convservation law which results in a partial differential equation known as the Liouville equation, for which a closed form solution is known.
- Chapter 9 provides conclusions and directions for future work.

2 Mathematical Preliminaries

This chapter provides background information on material that will be used throughout this work. Material covered includes the differential geometric or coordinate-free description of control systems, control system properties, nonholonomic systems, and chained forms.

2.1 Differential Geometric Description of Systems

The concepts of differential geometry are very important to the field of nonlinear control theory. General nonlinear systems may exist on manifolds other than $\mathbb{R}^n$ and their local behavior may be very different from their global behavior. This phenomenon is in contrast to linear systems existing in $\mathbb{R}^n$ whose global behavior is identical to the local behavior about a point.

As an example, consider the Möbius strip shown in Figure 2.1. Locally, the surface looks like a plane, but it is obviously not a plane when the entire surface is viewed in three dimensions. Given a vector v at any point, the vector maintains the same orientation as it travels a short distance on the curve c. However, if the vector travels along c around the entire strip, its orientation has changed when it returns to the original point, as shown by v'. The local behavior did not give any indication that the vector would reverse its direction. In this case, the local behavior is not a accurate predictor of the global behavior.

This section introduces the concepts of differential geometry. It provides the basic definitions and terminology used in this area. The material in this section is treated in the context of dynamical control systems in [21] and [44] and from a topological point of view in [37].

Definition 2.1. ***Topological Space:*** *A topological space is a set X together with $\mathcal{T}$, a collection of subsets of X, that satisfies the following:*

1. *$\mathcal{T}$ contains $\emptyset$ and X.*
2. *The union of any collection of sets in $\mathcal{T}$ is in $\mathcal{T}$.*
3. *The intersection of a finite collection of sets in $\mathcal{T}$ is in $\mathcal{T}$.*

P. Mellodge & P. Kachroo: Model Abstraction in Dynamical Systems, LNCIS 379, pp. 7–25, 2008.
springerlink.com

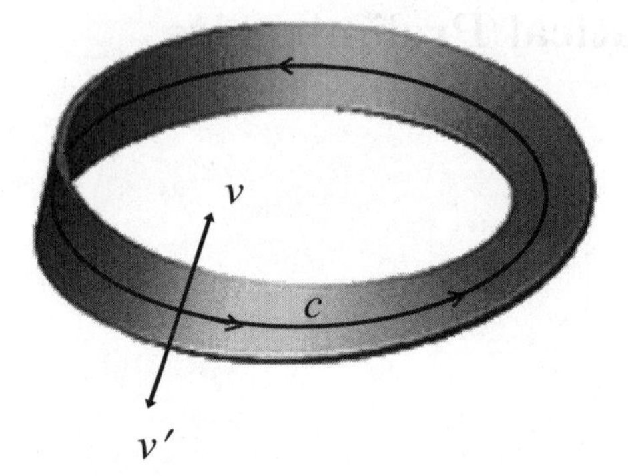

Fig. 2.1. A vector traveling along a curve on a Möbius strip

The topological space is denoted by $(X, \mathcal{T})$. Often the topological space is simply refered to as X when the topology $\mathcal{T}$ is apparent. The sets in $\mathcal{T}$ are called *open*. For any $x \in X$, an open set containing x is called a *neighborhood* of x.

Two immediate topologies on a set X are the discrete topology, in which all subsets of X are open, and the trivial topology, in which only $\emptyset$ and X are open. Most useful topologies fall somewhere between these two extremes.

Topologies on X can be described by a *basis*, a collection of sets $\mathcal{B} \subseteq \mathcal{T}$, such that every open set can be written as the union of elements of $\mathcal{B}$. However, the representation of open sets need not be unique.

The familiar concept of continuity can be defined topologically, without the use of a metric. The following definition relates open sets in the range of a function with open sets in its domain.

Definition 2.2. ***Continuity:*** *Let X and Y be topological spaces and suppose $f : X \to Y$. Then f is continuous if for every open set $O \subseteq Y$, f^{-1} is an open set in X.*

The existence of functions between topological spaces that preserve open sets is important to the study of topology. Such spaces, called *homeomorphic*, are topologically equivalent, and such functions are called *homeomorphisms*. The precise definition is as follows.

Definition 2.3. ***Homeomorphism:*** *Two topological spaces, X and Y, are said to be homeomorphic if there exists a function $f : X \to Y$ that is continuous and bijective (i.e., one-to-one and onto) and whose inverse f^{-1} is continuous. The function f is called a homeomorphism.*

Next we define a particular type of topological space that is well-behaved and eliminates a certain class of unusual spaces that are not useful in studying physical systems. The concept of a Hausdorff space is important because it ensures that the limit of a sequence of points, if it exists, is unique.

Definition 2.4. ***Hausdorff Space:*** *A topological space is called a Hausdorff space if, given two distinct points in the space, x_1 and x_2, there exist open sets O_1 and O_2 such that $x_1 \in O_1$, $x_2 \in O_2$, and $O_1 \cap O_2 = \emptyset$.*

The most familiar example of a Hausdorff space is likely the set of real numbers, $\mathbb{R}$. Almost as familiar are the Cartesian products of real numbers: $\mathbb{R}^2 = \mathbb{R} \times \mathbb{R}$ (the plane), $\mathbb{R}^3 = \mathbb{R} \times \mathbb{R} \times \mathbb{R}$ (the cube), etc. However, a space can be the Cartesian product of sets other than the real numbers. For example, consider the space consisting of the points (θ_1, θ_2) where both θ_1 and θ_2 range from 0 to 2π. Such a space can be represented by a sphere, as shown in Figure 2.2. While a sphere is not a plane, if a very small area of the sphere is viewed, it is very similar to a plane. This relationship between the small section of the sphere and the plane is the idea behind *manifolds.*

Definition 2.5. ***Manifold:*** *A set M is called a manifold if it is a Hausdorff space with a countable basis such that for any $p \in M$, there exists a neighborhood of p that is homeomorphic to an open set in $\mathbb{R}^n$. The dimension of M is n.*

A manifold is said to be *connected* if it is not the union of two disjoint open sets.

Manifolds can be characterized by *coordinate charts.* If U is an open subset of manifold M and ϕ is a homeomorphism between U and an open set in $\mathbb{R}^n$, then the pair (U, ϕ) is called a coordinate chart on M. Often ϕ is represented by an n-vector, $(\phi_1, ..., \phi_n)$. If $p \in U$, then the vector $(\phi_1(p), ..., \phi_n(p))$ represents the *local coordinates* of p. This is demonstrated in Figure 2.2 which shows that a sphere can be locally represented by a plane.

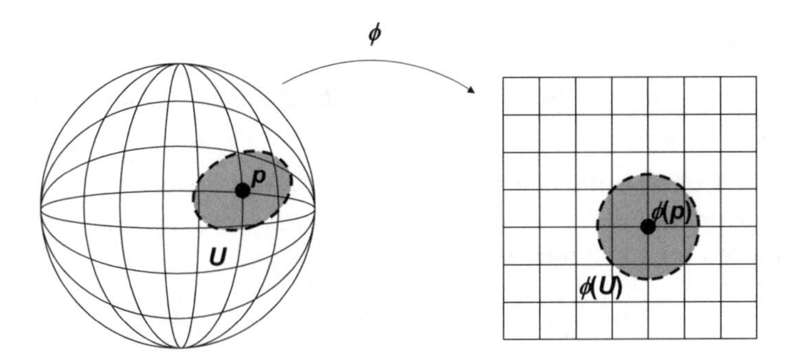

Fig. 2.2. The local representation of an open set on a sphere by a plane

Since these ideas will be applied to dynamical systems, differentiability on a manifold needs to be considered. The homeomorphisms play an important role because they allow the system characteristics (such as differentiation) to be studied in the local coordinates in $\mathbb{R}^n$ rather than the more complex manifold.

Coordinate charts (U, ϕ) and (V, ψ) are said to be C^∞-compatible if either of the following conditions holds:

1. $U \cap V = \emptyset$.
2. $U \cap V \neq \emptyset$ and the transformation given by $\psi \circ \phi^{-1}$ has a C^∞ inverse[1] given by $\phi \circ \psi^{-1}$.

The relationship between the coordinate charts is shown in Figure 2.3.

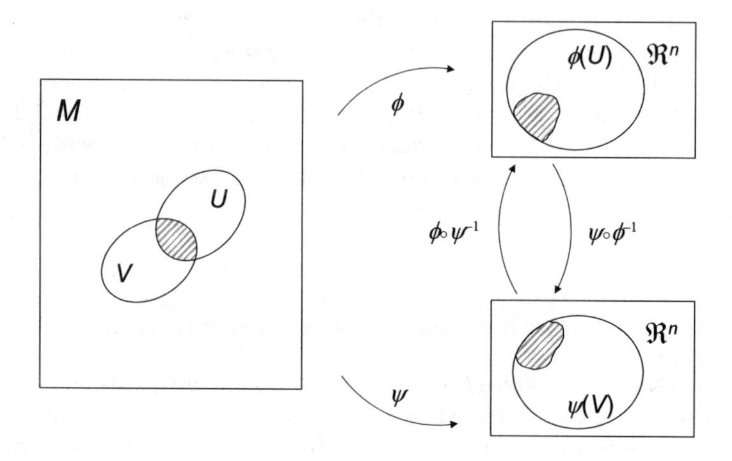

Fig. 2.3. C^∞-compatible coordinate charts (U, ϕ) and (V, ψ)

Definition 2.6. ***C^∞-atlas:*** *A collection of coordinate charts $\mathcal{U} = (U_\alpha, \phi_\alpha)_{\alpha \in \mathcal{A}}$ on manifold M is called a C^∞-atlas if the members of $\mathcal{U}$ are pairwise C^∞-compatible and $\cup_{\alpha \in \mathcal{A}} U_\alpha = M$. Furthermore, $\mathcal{U}$ is called complete if it is the largest such atlas.*

Definition 2.7. ***Smooth Manifold:*** *A smooth manifold is a manifold that has a complete C^∞-atlas.*

Example 2.8. Consider the manifold shown in Figure 2.4 given by $M = \mathbb{R} \times S^1$, i.e., the set of points (x, θ) where $x \in \mathbb{R}$ and $\theta \in [0, 2\pi]$. For any point $p = (x_0, \theta_0)$, define the open set containing p by

$$U_p = \left\{ (x, \theta) \mid |x - x_0| < 1 \text{ and } |\theta - \theta_0| < \frac{\pi}{4} \right\}.$$

Then define the homeomorphism $\phi_p : U_p \to \mathbb{R}^2$ by

$$\phi_p(x, \theta) = (x, \theta - \theta_0).$$

Let $\mathcal{U} = \{(U_p, \phi_p) : p \in M\}$. It is clear that $M = \cup_{p \in M} U_p$. For points in overlapping open sets, $p_1 = (x_1, \theta_1)$ and $p_2 = (x_2, \theta_2)$, the inverse mapping $\phi_{p_2} \circ \phi_{p_1}^{-1}$ is C^∞ because for any point $(y, \alpha) \in \phi_{p_2}(U_{p_2})$,

$$\phi_{p_2} \circ \phi_{p_1}^{-1}(y, \alpha) = \phi_{p_2}(y, \alpha + \theta_1) = (y, \alpha + \theta_1 - \theta_2).$$

[1] A function is C^k if partial derivatives of k orders exist and are continuous.

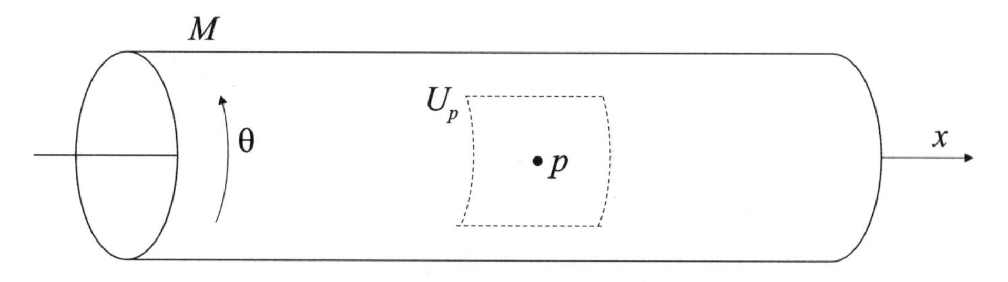

Fig. 2.4. The manifold for Example 2.8

Thus the coordinate charts in $\mathcal{U}$ are C^∞-compatible and $\mathcal{U}$ is a C^∞-atlas. Denote by $\mathcal{U}^*$ the set of all coordinate charts that are C^∞-compatible with elements of $\mathcal{U}$. Then $\mathcal{U}^*$ is a complete C^∞-atlas on M.

By extending this example to manifolds such as $\mathbb{R}\times\mathbb{R}\times S^1$ and $\mathbb{R}\times\mathbb{R}\times S^1\times S^1$, it is shown that the manifolds on which the unicycle and car models for the robot reside are smooth manifolds. Therefore, the theory developed for smooth manifolds in this chapter are valid for the models given in Chapter 3. From this point forward, it will be assumed that manifolds are smooth.

Diffeomorphisms are the next concept to be introduced. These are similar to coordinate transformations and are useful in the study of nonlinear systems. For example, diffeomorphisms are used for feedback linearization.

Definition 2.9. ***Diffeomorphism:*** *A map $f : M \to N$ between smooth manifolds is a diffeomorphism if f is bijective and both f and f^{-1} are smooth.*

The *rank* of $f : M \to N$ at a point $p \in M$ is given by the rank of the Jacobian matrix evaluated at $x = \phi(p)$ as follows

$$\mathrm{rank}_p(f) = \begin{pmatrix} \frac{\partial f_1}{\partial x_1} & \cdots & \frac{\partial f_1}{\partial x_n} \\ \vdots & \ddots & \vdots \\ \frac{\partial f_m}{\partial x_1} & \cdots & \frac{\partial f_m}{\partial x_n} \end{pmatrix}. \tag{2.1}$$

The rank is independent of the choice of local coordinates. If for all $p \in M$, $\mathrm{rank}_p(f) = \dim(M)$, then f is an *immersion.* If for all $p \in M$, $\mathrm{rank}_p(f) = \dim(N)$, then f is a *submersion.*

In the following definition, the set of all smooth real-valued functions in a neighborhood of $p \in M$ is denoted by $C^\infty(p)$.

Definition 2.10. ***Tangent Vector:*** *A tangent vector at $p \in M$ is a map $X_p : C^\infty(p) \to \mathbb{R}$ that satisfies the following for all $\lambda, \gamma \in C^\infty(p)$ and $\alpha, \beta \in \mathbb{R}$:*

1. *Linearity: $X_p(\alpha\lambda + \beta\gamma) = \alpha X_p(\lambda) + \beta X_p(\gamma)$,*
2. *Product Rule: $X_p(\lambda\gamma) = X_p(\lambda)\gamma(p) + \lambda(p)X_p(\gamma)$.*

The *tangent space to p at M,* denoted by T_pM, is the set of all tangent vectors at p. The *tangent bundle* of M is defined as $TM = \cup_{p\in M} T_pM$.

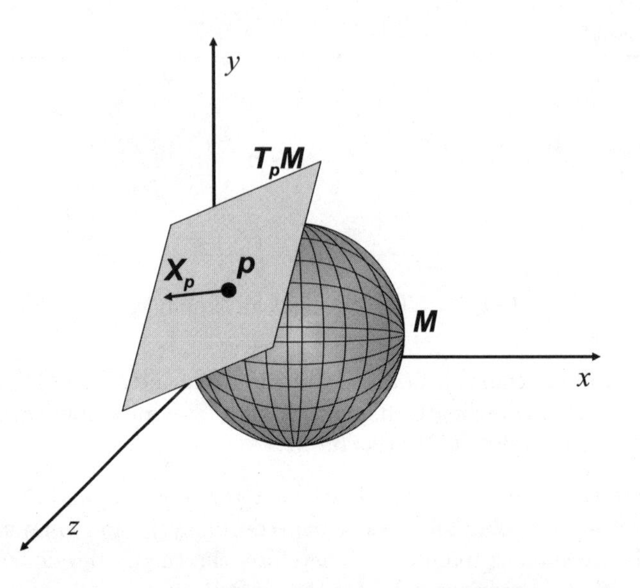

Fig. 2.5. The geometric interpretation of a tangent space

This representation can be associated with the familiar notion of a tangent space as the n-dimensional plane tangent to M at p, as shown in Figure 2.5. Since M is n-dimensional, a tangent vector at p can be decomposed into n tangent vectors, $\left(\frac{\partial}{\partial \phi_1}\right)_p, \cdots, \left(\frac{\partial}{\partial \phi_n}\right)_p$ by defining

$$\left(\frac{\partial}{\partial \phi_i}\right)_p (\lambda) = \left[\frac{\partial(\lambda \circ \phi^{-1})}{\partial x_i}\right]_{x=\phi(p)}. \tag{2.2}$$

This view provides a more intuitive characterization of tangent vectors as follows. Given $p \in M$ and coordinate chart (U, ϕ), the tangent vectors $\left(\frac{\partial}{\partial \phi_1}\right)_p, ..., \left(\frac{\partial}{\partial \phi_n}\right)_p$ form a basis for T_pM. In other words, the tangent vector X_p can be represented by

$$X_p = \sum_{i=1}^{n} \alpha_i \left(\frac{\partial}{\partial \phi_i}\right)_p \tag{2.3}$$

where the α_i's are real numbers. Thus,

$$X_p(\lambda) = \sum_{i=1}^{n} \alpha_i \left(\frac{\partial \lambda}{\partial \phi_i}\right)_p. \tag{2.4}$$

In the case that $M = \mathbb{R}^n$, $p = (x_1, ..., x_n)$, and $\phi(p) = p$, this reduces to

$$X_p(\lambda) = \sum_{i=1}^{n} \alpha_i \left(\frac{\partial \lambda}{\partial x_i}\right)_p. \tag{2.5}$$

Thus the value $X_p(\lambda)$ can be interpreted as the derivative of the smooth function λ in the direction of X_p.

Definition 2.11. ***Vector Field:*** *A vector field on n-dimensional manifold M at p is a mapping $f : M \to T_pM$. Furthermore, f is a smooth vector field if there exists a coordinate chart (U, ϕ) about p and $\mathbb{R}$-valued functions, $f_1, ..., f_n$ on U, such that*

$$f(q) = \sum_{k=1}^{n} f_k(q) \left(\frac{\partial}{\partial \phi_k} \right)_q \tag{2.6}$$

for all $q \in U$.

The vector field is a map that associates a tangent vector with each point of M, as shown in Figure 2.6. If a point continuously moves in the direction of a vector field for an interval of time, the resulting collection of points is an *integral curve* defined formally as follows.

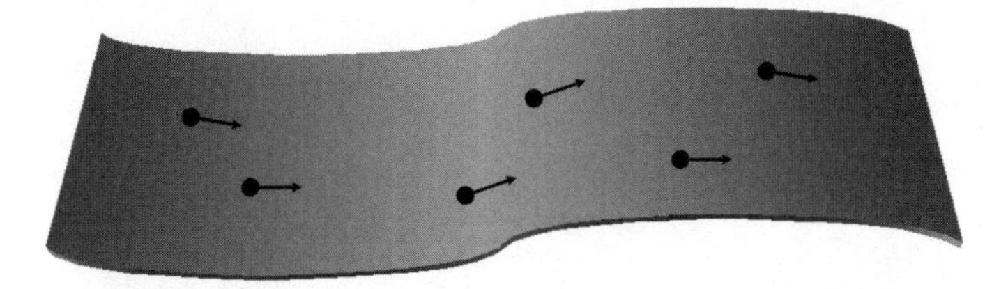

Fig. 2.6. Vector fields on a manifold

Definition 2.12. ***Integral Curve:*** *Given a vector field f on M, a smooth curve $c : (t_1, t_2) \to M$ is called an integral curve if*

$$\dot{c}(t) = f(c(t)) \tag{2.7}$$

for all $t \in (t_1, t_2)$.

An integral curve on M is a curve that follows a given vector field at each point. This behavior is illustrated in Figure 2.7.

Definition 2.13. ***Lie Algebra:*** *A vector space V together with the binary operation $[\cdot,\cdot] : V \times V \to V$ is called a Lie algebra if $[\cdot,\cdot]$ satisfies the following for all $v, w, z \in V$ and $\alpha, \beta \in \mathbb{R}$.*

1. *Bilinearity:* $[\alpha v + \beta w, z] = \alpha[v, z] + \beta[w, z]$.
2. *Anti-symmetry:* $[v, w] = -[w, v]$.
3. *Jacobi-identity:* $[v, [w, z]] + [w, [z, v]] + [z, [v, w]] = 0$.

The derivative of a smooth function λ in the direction of the vector field f is given by

$$L_f\lambda(p) = (f(p))(\lambda) \tag{2.8}$$

and is the definition of a *Lie derivative.*

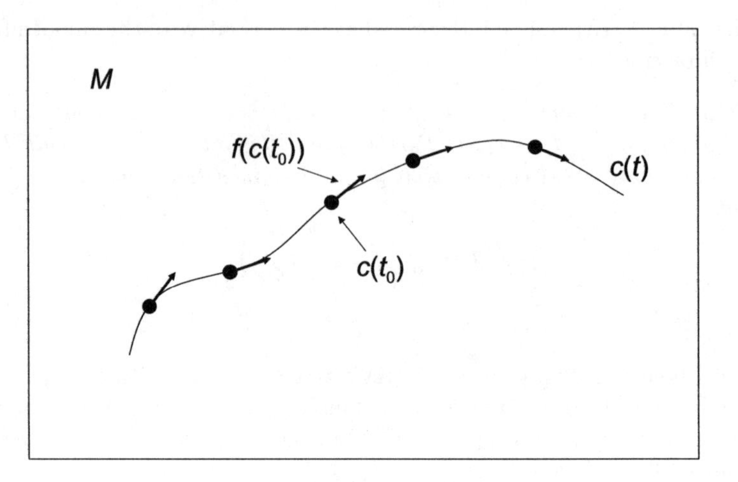

Fig. 2.7. An integral curve on manifold M

Example 2.14. Consider the vector field $f : \mathbb{R}^2 \to \mathbb{R}^2$ given by $f(x_1, x_2) = \begin{bmatrix} x_1^2 + 1 \\ x_1 + x_2 \end{bmatrix}$ and the smooth function λ given by $\lambda(x_1, x_2) = x_1^2 + x_2^2 + 10$. Then

$$L_f\lambda(x_1, x_2) = (\nabla\lambda f)(x_1, x_2) = [2x_1 \; 2x_2] \begin{bmatrix} x_1^2 + 1 \\ x_1 + x_2 \end{bmatrix}. \tag{2.9}$$

For the point $(x_1, x_2) = (2, 1)$, $f(2, 1) = \begin{bmatrix} 5 \\ 3 \end{bmatrix}$ and $L_f\lambda(2, 1) = [4 \; 2] \begin{bmatrix} 5 \\ 3 \end{bmatrix} = 26$. These values indicate that the slope of the tangent of λ at the point $(2, 1)$ in the direction $\begin{bmatrix} 5 \\ 3 \end{bmatrix}$ is 26 (see Figure 2.8).

Given two vector fields, f and g, a new vector field can be defined by

$$([f, g](p))(\lambda) = (L_f L_g \lambda)(p) - (L_g L_f \lambda)(p) \tag{2.10}$$

and $[f, g]$ is called the *Lie bracket* of f and g. A vector space with the Lie bracket as the binary operation is a Lie algebra. Denote by $V^\infty(M)$ the space of C^∞ vector fields on M. Together with the Lie bracket operation, $V^\infty(M)$ is an algebra.

Example 2.15. Consider the vector fields on $\mathbb{R}^2$ given by $f = \begin{bmatrix} \sin x_1 x_2 \\ x_2 \end{bmatrix}$ and $g = \begin{bmatrix} x_1 \\ 0 \end{bmatrix}$.

Then the Lie bracket is given by

$$[f,g] = \nabla g f - \nabla f g \tag{2.11}$$

$$= \begin{bmatrix} 1 & 0 \\ 0 & 0 \end{bmatrix} \begin{bmatrix} \sin x_1 x_2 \\ x_2 \end{bmatrix} - \begin{bmatrix} x_2 \cos x_1 x_2 & x_1 \cos x_1 x_2 \\ 0 & 1 \end{bmatrix} \begin{bmatrix} x_1 \\ 0 \end{bmatrix} \tag{2.12}$$

$$= \begin{bmatrix} \sin x_1 x_2 - x_1 x_2 \cos x_1 x_2 \\ 0 \end{bmatrix}. \tag{2.13}$$

Definition 2.16. ***Distribution:*** *A distribution on M is a mapping $\Delta : M \rightarrow TM$ that assigns to each $p \in M$ a subspace of the tangent space T_pM. Furthermore, Δ is called a smooth distribution if for every $p \in M$ there exists a neighborhood U and a collection of smooth vector fields f_i, $i = 1...m$, such that for all $q \in U$*

$$\Delta(q) = \text{span}(f_i(q) \,|\, i = 1, ..., m) \tag{2.14}$$

The dimension of $\Delta(q)$ is m.

A distribution Δ is called *involutive* if, given any two vector fields $f, g \in \Delta$, the Lie bracket $[f, g]$ is also in Δ.

A distribution Δ defined on an open set U is called *nonsingular* if there exists some integer m such that for all $p \in U$

$$\dim(\Delta(p)) = m. \tag{2.15}$$

A nonsingular r-dimensional distribution Δ, defined on an open subset U of $\mathbb{R}^n$, is called *completely integrable* if for each $p \in U$ there exists a neighborhood U_p of p and $n - r$ real-valued smooth functions, $\lambda_1, ..., \lambda_{n-r}$, on U such that

$$\text{span}(d\lambda_1, ..., d\lambda_{n-r}) = \{w \in (\mathbb{R}^n)^* \mid \langle w, v \rangle = 0 \text{ for all } v \in \Delta(p)\}. \tag{2.16}$$

In this definition, $(\mathbb{R}^n)^*$ is the *dual space* of $\mathbb{R}^n$, i.e., the space of all bounded linear functionals mapping $\mathbb{R}^n$ to the complex numbers. Also, $v = [v_1\, v_2 \cdots v_n]^T$, $w = [w_1\, w_2 \cdots w_n]$, and $\langle w, v \rangle = \sum_{i=1}^{n} w_i v_i$. The righthand side of (2.16) is called the *annilhilator* and is denoted by $\Delta^{\perp}$.

Given a distribution Δ, the *filtration* generated by Δ is a sequence $(\Delta_1, \Delta_2, ...)$ where

$$\begin{aligned} \Delta_1 &= \Delta \\ \Delta_{i+1} &= \Delta_i + \text{span}\{[f,g] \,|\, f \in \Delta_1, g \in \Delta_i\} \end{aligned} \tag{2.17}$$

The following theorem is a fundamental result which relates to, among other things, the existence of solutions of partial differential equations and the controllability of a system. It gives conditions for a system to be broken down into reachable and unreachable parts, an important step in determining the controllability of a system. The proof can be found in [21].

Theorem 2.17. ***Frobenius:*** *A nonsingular distribution is completely integrable if and only if it is involutive.*

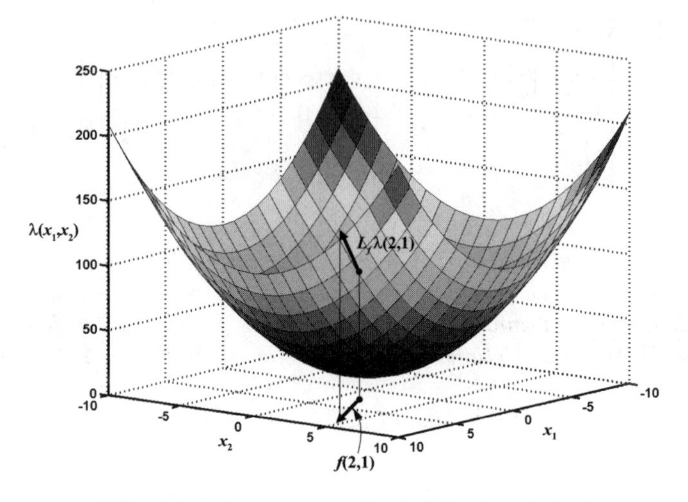

Fig. 2.8. The interpretation of the Lie derivative

The following example, from [53], shows how Frobenius' Theorem can be used to determine the existence of solutions to a set of partial differential equations.

Example 2.18. Consider the following partial differential equations.

$$4x_3\frac{\partial h}{\partial x_1} - \frac{\partial h}{\partial x_2} = 0 \tag{2.18}$$

$$-x_1\frac{\partial h}{\partial x_1} + (x_3^2 - 3x_2)\frac{\partial h}{\partial x_2} + 2x_3\frac{\partial h}{\partial x_3} = 0 \tag{2.19}$$

The vector fields are given by

$$f = \begin{bmatrix} 4x_3 \\ -1 \\ 0 \end{bmatrix} \text{ and } g = \begin{bmatrix} -x_1 \\ x_3^2 - 3x_2 \\ 2x_3 \end{bmatrix}. \tag{2.20}$$

Then $\Delta = \text{span}(f, g)$ and

$$[f, g] = \begin{bmatrix} -1 & 0 & 0 \\ 0 & -3 & 2x_3 \\ 0 & 0 & 2 \end{bmatrix} \begin{bmatrix} 4x_3 \\ -1 \\ 0 \end{bmatrix} - \begin{bmatrix} 0 & 0 & 4 \\ 0 & 0 & 0 \\ 0 & 0 & 0 \end{bmatrix} \begin{bmatrix} -x_1 \\ x_3^2 - 3x_2 \\ 2x_3 \end{bmatrix} = \begin{bmatrix} -12x_3 \\ 3 \\ 0 \end{bmatrix}. \tag{2.21}$$

This calculation shows that $[f, g] = -3f$, so that $[f, g] \in \Delta$ and $r = 2$. Thus Δ is involutive and, by Frobenius' Theorem, completely integrable. Since $n - r = 1$, there exists a single real-valued smooth function h such that

$$\frac{\partial h}{\partial x} f(x) = 0 \tag{2.22}$$

$$\frac{\partial h}{\partial x} g(x) = 0. \tag{2.23}$$

The ideas in this section are important in the study of control systems of the form $\dot{x} = f(t, x(t), u(t))$. It is often the case that the control system's state vector is not in $\mathbb{R}^n$ but rather in a more general manifold. The vector fields of the system are given by $f(t, x(t), u(t))$ and the integral curves define the trajectories of the system. The tools in this section provide a means to study the system's global behavior, the most important of which is controllability.

2.2 Control System Properties

In the study of control systems, controllability is particularly important. Often, the first step in control design is to determine which states are controllable. If none of the states are controllable, then no controller can be designed to achieve the desired result. If only some of the states are controllable, then all may not be lost since the control objective may involve only those states.

We are generally concerned with systems of the form

$$\dot{x}(t) = f(t, x(t), u(t)) \tag{2.24}$$

where $x(t) = [x_1(t), ..., x_n(t)]^T \in M$ is the state vector of the system on some manifold M, $x_1, ..., x_n$ are the state variables (or states), and $u(t) = [u_1(t), ..., u_m(t)]^T \in U$, is the control input to the system that belongs to some set of admissible control inputs U. This is the most general form of a dynamic system since it is both time-varying and nonlinear.

Definition 2.19. ***Controllability:*** *The system given by (2.24) is controllable at $x_0 \in M$ if, given $x'_0 \in M$, there exists some $u(t) \in U$ and finite T such that $x(0) = x_0$ and $x(T) = x'_0$. The system is controllable if it is controllable at every $x_0 \in M$.*

In other words, a system is controllable if there exists an input that can move the system between any two points in finite time.

For linear time-invariant systems of the form

$$\dot{x}(t) = Ax(t) + Bu(t) \tag{2.25}$$

where $x(t) \in \mathbb{R}^n$, $u(t) \in \mathbb{R}^m$, A is an $n \times n$ matrix, and B is an $n \times m$ matrix, controllability can be determined by the familiar *rank condition* as follows.

Theorem 2.20. *The system given in (2.25) is controllable if and only if*

$$\text{rank}(B \; AB \; A^2B \; \cdots \; A^{n-1}B) = n. \tag{2.26}$$

The idea behind the proof is that a linear system is controllable if it can be driven to zero from any initial state in finite time. When the solution to (2.25) is obtained and set to zero, it can be expressed as

$$\xi = \begin{bmatrix} B \; AB \; A^2B \; \cdots \; A^{n-1}B \end{bmatrix} \begin{bmatrix} \Gamma_n \\ \vdots \\ \Gamma_1 \end{bmatrix} \tag{2.27}$$

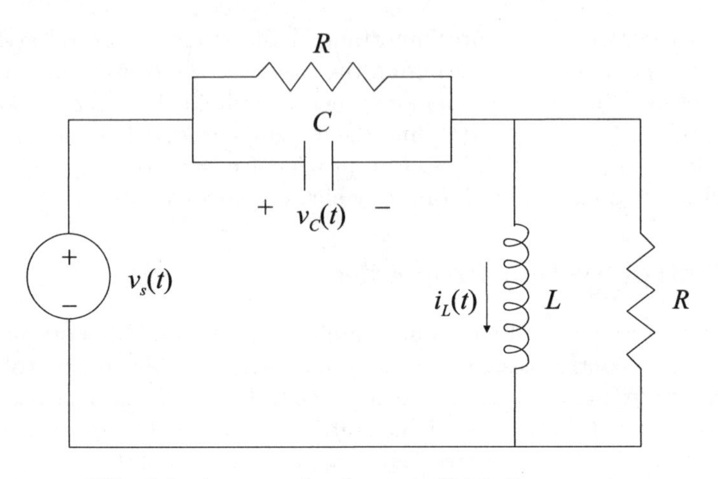

Fig. 2.9. An example of a controllable linear system

where ξ is arbitrary and $\Gamma_1, ..., \Gamma_n$ are obtained using the Cayley-Hamilton expansion of $e^{A(T-t)}$. Since the solution to this matrix equation must exist for any ξ, then $[B \; AB \; A^2B \; \cdots \; A^{n-1}B]$ is required to have full rank. Complete details of this proof can be found in [4].

The following example from [4] illustrates controllability analysis for a linear system.

Example 2.21. Consider the circuit shown in Figure 2.9. The state variables are v_C, the voltage across the capacitor, and i_L, the current through the inductor. The input to the system is v_s. The state equations are given by

$$\begin{bmatrix} \dot{v}_C(t) \\ \dot{i}_L(t) \end{bmatrix} = \begin{bmatrix} -\frac{2}{RC} & \frac{1}{C} \\ -\frac{1}{L} & 0 \end{bmatrix} \begin{bmatrix} v_C(t) \\ i_L(t) \end{bmatrix} + \begin{bmatrix} \frac{1}{RC} \\ \frac{1}{L} \end{bmatrix} v_s(t). \tag{2.28}$$

The rank condition is then equivalent to checking the determinant of the controllability matrix.

$$|B \; AB| = \begin{vmatrix} \frac{1}{RC} & -\frac{2}{R^2C^2} + \frac{1}{LC} \\ \frac{1}{L} & -\frac{1}{RLC} \end{vmatrix} = \frac{1}{R^2LC^2} - \frac{1}{L^2C} \tag{2.29}$$

When the determinant is nonzero, the rank of the matrix is 2. So the system is controllable for $R \neq \sqrt{\frac{L}{C}}$.

Now consider the more general smooth affine nonlinear systems of the form

$$\dot{x}(t) = f(x) + \sum_{i=1}^{m} g_i(x)u_i \tag{2.30}$$

where

$$\mathcal{F} = \{f + \sum_{i=1}^{m} g_i u_i \,|\, (u_1, ..., u_m) \in U\} \tag{2.31}$$

contains $f, g_1, ..., g_m$.

Before investigating the controllability of this system, the weaker notion of accessibility is discussed.

Let $R_T^V(x_0)$ be the *reachable set* from x_0 in time $T > 0$ defined as follows.

$$\begin{aligned} R_T^V(x_0) = \{x \in M \mid &\text{there exists } u \in U \text{ such that } x(0) = x_0, \\ &x(t) \in V \text{ for all } t \in [0, T], \text{ and } x(T) = x\} \end{aligned} \tag{2.32}$$

Definition 2.22. ***Locally Accessible:*** *The system given by (2.24) is locally accessible at x_0 if for all neighborhoods V of x_0 and all $T > 0$, $R_V^T(x_0)$ contains an open subset of M. If the system is locally accessible at x_0 for all $x_0 \in M$, then it is called locally accessible.*

Denote by $\mathcal{C}$ the smallest subalgebra of V^∞ that contains $f, g_1, ..., g_m$. This subalgebra $\mathcal{C}$ is known as the *accessibility algebra.* Every vector field in $\mathcal{C}$ can be represented by a linear combination of iterated Lie brackets of the form

$$[X_k, [X_{k-1}, [..., [X_2, X_1]...]]] \tag{2.33}$$

where $k = 1, 2, ...$ and $X_i \in \{f, g_1, ..., g_m\}$ for $i \in [1, k]$. Also, denote by $C(x)$ the span of all vector fields in $\mathcal{C}$.

The local accessibility of the system in (2.30) can be determined by the dimension of $C(x)$ as given by the following theorem. This theorem, whose proof is given in [44], is known as the *accessibility rank condition.* In effect, the theorem states that if there are n independent vector fields in C, then the system can locally move in all directions on M.

Theorem 2.23. *If* $\dim(C(x)) = n$ *for every $x \in M$, then (2.30) is locally accessible.*

The next theorem, also proven in [44], provides a useful characterization of controllability of a driftless system in terms of $C(x)$. It takes the result of the previous theorem and extends it to give sufficient conditions for a system to be controllable. The additional requirement is that the vector fields be symmetric, i.e., if u is an admissible control input, then so too is $-u$.

Theorem 2.24. *Suppose $f = 0$ in (2.30) and $X \in \mathcal{F}$ implies that $-X \in \mathcal{F}$. Then if* $\dim(C(x)) = n$ *for all $x \in M$ and M is connected, then (2.30) is controllable.*

An equivalent statement of the above theorem is *Chow's Theorem.* It states the controllability result explicitly in terms of the Lie brackets of the vector fields. A proof is given in [6].

Theorem 2.25. ***Chow's Theorem:*** *Suppose (2.30) is a symmetric control system with $f = 0$ and M is a connected manifold. If for each $p \in M$ the vector fields $g_1, g_2, ..., g_m$ and their iterated Lie brackets span T_pM, then the system is controllable.*

The condition that the vector fields and their iterated Lie brackets span the tangent space is known as *Chow's Condition* or the *Lie Algebra Rank Condition* (LARC). An application of Chow's theorem is given in the following example. The model used in the example for a unicycle as described in the next chapter.

Example 2.26. Consider the following driftless system.

$$\begin{bmatrix} \dot{x}_1 \\ \dot{x}_2 \\ \dot{x}_3 \end{bmatrix} = \begin{bmatrix} \cos x_3 \\ \sin x_3 \\ 0 \end{bmatrix} u_1 + \begin{bmatrix} 0 \\ 0 \\ 1 \end{bmatrix} u_2 \tag{2.34}$$

where $[x_1, x_2, x_3]^T$ is in the connected set $\mathbb{R} \times \mathbb{R} \times [0, 2\pi]$ and $u_1, u_2 \in \mathbb{R}$. The controllability of this system is determined by checking the rank of the matrix

$$[g_1 \ g_2 \ [g_1, g_2] \ ...] \tag{2.35}$$

where

$$g_1 = \begin{bmatrix} \cos x_3 \\ \sin x_3 \\ 0 \end{bmatrix}, \ g_2 = \begin{bmatrix} 0 \\ 0 \\ 1 \end{bmatrix}. \tag{2.36}$$

The Lie bracket $[g_1, g_2]$ is given by

$$[g_1, g_2] = \nabla g_2 g_1 - \nabla g_1 g_2 = - \begin{bmatrix} 0 \ 0 & -\sin x_3 \\ 0 \ 0 & \cos x_3 \\ 0 \ 0 & 0 \end{bmatrix} \begin{bmatrix} 0 \\ 0 \\ 1 \end{bmatrix} = \begin{bmatrix} \sin x_3 \\ -\cos x_3 \\ 0 \end{bmatrix}. \tag{2.37}$$

Since $[g_1, g_2]$ is independent of g_1 and g_2, the dimension of $[g_1 \ g_2 \ [g_1, g_2]]$ is 3. Thus the system is controllable.

2.3 Nonholonomic Systems

Nonholonomic systems are important in the study of robotics. Most robotic systems exhibit nonholonomic behavior under normal operating conditions and this behavior leads to simplified models to which the theory developed in the previous sections readily applies.

As described in [60], the study of nonholonomic systems has its origins in the study of classical mechanics. While such systems were studied by Euler, a clear understanding of nonholonomic system structures in the context of mechanics was obtained by Hertz in the 1890s. The theory of nonholonomic systems has connections with other areas as well: electromechanics, as studied by Maxwell; thermodynamics, as studied by Gibbs and Caratheodory; and quantum mechanics, as studied by Dirac.

The word *holonomic*, from the Greek words ὀλος, meaning "entire," and νομος, meaning "law," was coined by Hertz and means "universal," or "integrable." *Nonholonomic* therefore means "nonintegrable."

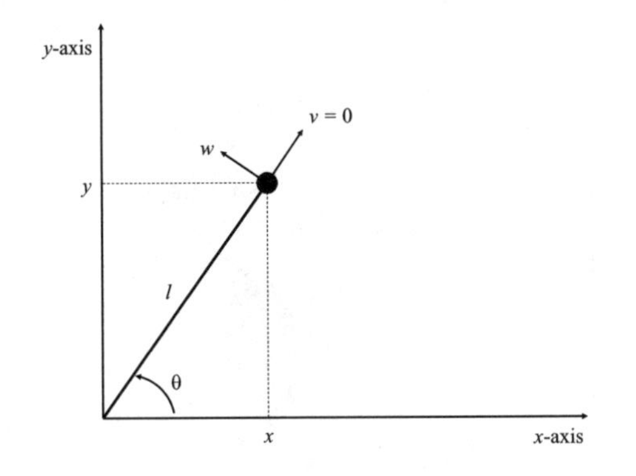

Fig. 2.10. The geometric and kinematic constraints of a robotic arm

For a mechanical system, we can consider two kinds of constraints: geometric and kinematic. Geometric constraints are position restrictions while kinematic constraints are velocity restrictions. Geometric constraints always impose velocity constraints on a system, but the converse is not true [43].

As an example, consider a simple planar robotic arm as shown in Figure 2.10. The position of the end of the arm is given by (x, y). Assuming the arm is rigid, it has some fixed length l. Then the geometric constraint is given by $x^2 + y^2 = l^2$. This constraint leads directly to velocity restrictions since the endpoint cannot move in the direction of v, but must move in the direction of w. The kinematic constraints can then be expressed as follows.

$$\dot{x} + l\dot{\theta}\sin\theta = 0 \tag{2.38}$$

$$\dot{y} - l\dot{\theta}\cos\theta = 0 \tag{2.39}$$

Now consider a unicycle, as shown in Figure 2.11, moving in the (x, y)-plane with the no-slip condition, i.e., the unicycle can only move in the direction of rolling. This condition means that there can be no velocity in the direction of w and so it must all be in the direction of v. So the following kinematic constraint must be satified at all times.

$$\dot{x}\sin\theta - \dot{y}\cos\theta = 0. \tag{2.40}$$

Or equivalently,

$$\begin{bmatrix} \sin\theta & -\cos\theta & 0 \end{bmatrix} \begin{bmatrix} \dot{x} \\ \dot{y} \\ \dot{\theta} \end{bmatrix} = 0. \tag{2.41}$$

However, this kinematic constraint does not lead to position constraints on the unicycle system. We know intuitively that the unicycle can go to any point in the plane[2]. For example, in Figure 2.12 the unicycle is able to move from position

[2] That is, in the absence of obstacles. If obstacles are present, they introduce geometric constraints that are not imposed by the kinematic constraints.

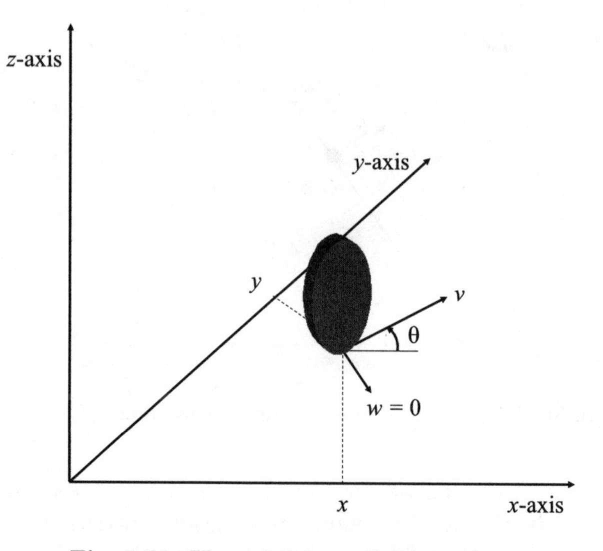

Fig. 2.11. The unicycle moving in a plane

A to position B even though its movement is restricted in that direction. For instance, it can simply move forward, rotate, move backward, and rotate again to achieve position B.

If we describe the system using generalized coordinates $(q_1, ..., q_n)$, then the geometric constraints can be expressed by

$$\begin{aligned} f_1(q_1, ..., q_n, t) &= 0 \\ &\vdots \\ f_m(q_1, ..., q_n, t) &= 0. \end{aligned} \tag{2.42}$$

The kinematic constraints are of the form

$$\begin{aligned} f_1(q_1, ..., q_n, \dot{q}_1, ..., \dot{q}_n, t) &= 0 \\ &\vdots \\ f_m(q_1, ..., q_n, \dot{q}_1, ..., \dot{q}_n, t) &= 0. \end{aligned} \tag{2.43}$$

A specific type of kinematic constraint is the *Pfaffian* constraint that can be expressed as $C(q)\dot{q} = 0$, where q is a vector of generalized coordinates and $C(q)$ is the Pfaffian constraint matrix. The size of the Pfaffian constraint matrix is $m \times n$, where m is the number of kinematic constraint equations and n is the size of the state space. The constraints in (2.41) are of this form. The nullspace of the matrix $C(q)$ gives the admissible velocities of the system.

If the differential equations in (2.43) are integrable, then then kinematic constraints are said to be *integrable.* Otherwise they are *nonintegrable.* Kinematic constraints that are integrable can be reduced to geometric constraints. For

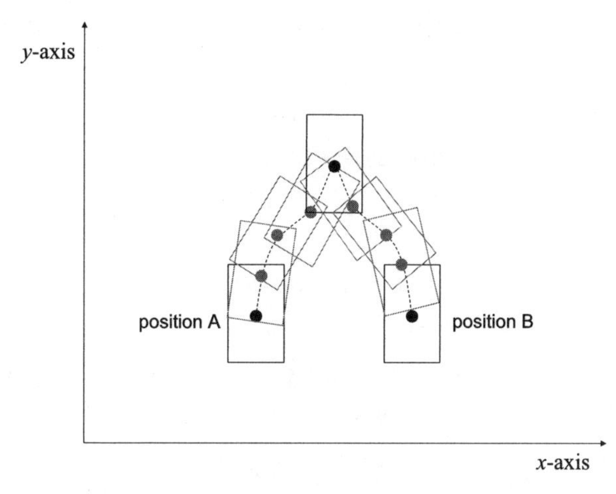

Fig. 2.12. The unicycle can move from position A to position B even though its velocity is restricted

example, in the case of the robotic arm above, the position constraints could be obtained by integration of the velocity restrictions.

Kinematic constraints that are nonintegrable are not in general equivalent to geometric constraints. This behavior is illustrated by the unicycle. The velocity restrictions cannot be integrated into position restrictions. While it may not be obvious that (2.41) is not integrable, intuitively we know that there are no geometric constraints. Therefore the kinematic constraints must be nonintegrable.

A *nonholonomic* system is one in which the kinematic constraints are nonintegrable and cannot be reduced to geometric constraints. In other words, a nonholonomic system is one in which the movement is restricted locally, but not globally. Usually, nonintegrable kinematic constraints are present when a system has fewer control inputs than states. In particular, it is necessary that all of the kinematic constraints are nonintegrable. A constraint may be nonintegrable on its own, but when taken together with the others, it may become integrable.

Controllability and nonholonomy are closely related. If a system with kinematic constraints is controllable, then it is nonholonomic. This is because controllability implies that the system can attain any position in its state space. In general, it is easier to determine whether a system is controllable than it is to determine whether its kinematic constraints are nonintegrable.

The *degree of nonholonomy at a point*, d_p, is the minimum length of the Lie bracket required to span the tangent space at that point. The degree of nonholonomy for the system, d, is the upper bound of degrees of nonholonomy for each point of the system's space, i.e., $d = \sup_{p \in M} d_p$. Equivalently, the degree of nonholonomy is the smallest integer k such that $\dim(\Delta_{k+1}) = \dim(\Delta_k)$.

Example 2.27. Consider again the unicycle system given in (2.34). For the given g_1 and g_2,

$$\begin{aligned} \Delta_1 &= \text{span}\{g_1, g_2\} \\ \dim(\Delta_1) &= 2 \end{aligned} \tag{2.44}$$

and since $[g_1, g_2]$ is not in Δ,

$$\begin{aligned} \Delta_2 &= \text{span}\{g_1, g_2, [g_1, g_2]\} \\ \dim(\Delta_2) &= 3. \end{aligned} \tag{2.45}$$

For $k > 2$, $\dim(\Delta_k) = 3$ because $\dim(\Delta_k) \leq \dim(\Delta_{k+1}) \leq 3$. Therefore the degree of nonholonomy for the unicycle is 2.

2.4 Chained Form

There is an important class of nonholonomic systems that can be converted into a canonical form known as chained form. As is the case with many other canonical forms, the controllability of chained form systems has been shown and controllers have been developed. For example, see [52] and [38].

The $(2, n)$ single-chain form has two inputs and n states and is given by [39]

$$\begin{aligned} \dot{x}_1 &= u_1 \\ \dot{x}_2 &= u_2 \\ \dot{x}_3 &= x_2 u_1 \\ &\vdots \\ \dot{x}_n &= x_{n-1} u_1. \end{aligned} \tag{2.46}$$

It can be shown that that all Lie brackets of length $n-2$ or greater are zero. This property is known as *nilpotency.* For the case where $n = 3$, the chained system reduces to

$$\begin{aligned} \dot{x}_1 &= u_1 \\ \dot{x}_2 &= u_2 \\ \dot{x}_3 &= x_2 u_1. \end{aligned}$$

The controllability of this system is determined by checking the rank of the matrix

$$[g_1 \; g_2 \; [g_1, g_2] \; ...] \tag{2.47}$$

where

$$g_1 = \begin{bmatrix} 1 \\ 0 \\ x_2 \end{bmatrix}, \; g_2 = \begin{bmatrix} 0 \\ 1 \\ 0 \end{bmatrix}. \tag{2.48}$$

Since

$$[g_1, g_2] = \begin{bmatrix} 0 \\ 0 \\ 1 \end{bmatrix} \tag{2.49}$$

is linearly independent of g_1 and g_2 for all values of x, the rank of the matrix is 3 and the system is controllable.

In [38], sufficient conditions are given for the conversion of a system into chained form. Later in Chapter 3, the transformations will be given to convert the unicycle and car models into chained form. Controllers are developed using this form.

3 Kinematic Modeling and Control

This chapter describes the kinematic modeling of a car-like mobile robot. Kinematic modeling is often used because of its simplicity and accuracy in predicting the car's behavior under normal driving conditions. This type of modeling uses the nonholonomic constraints of the system as described in Section 2.3.

After the modeling is discussed, the control design from [34] is described. This controller uses the path's curvature as a parameter. Since the curvature must be determined, two estimation methods are described and simulation results are presented.

3.1 Modeling Approach

The control of a mobile robot can be viewed as a hierarchical system of three controllers: the motion planner, the motion controller (which performs point-to-point stabilization, path following, or trajectory tracking), and the actuator driver. This structure is illustrated in Figure 3.1.

At the top of the hierarchy is the motion planner. The motion planner determines what path and what velocity profile the robot is to follow. At least part of this motion planning must be done off-line to determine the ultimate goal of the robot's motion. For this work, it is assumed that the path and velocity are determined beforehand (e.g., a line to follow is painted on the road and the robot must maintain a safe speed at all times). However, on-line motion planners are available so that the robot can modify its movement based upon changing environmental factors. See [28] for example.

Once the mobile robot knows where it must go, the controller at the next level takes on the task of making it get there. At this level, the actual position and velocity of the robot are measured and compared to the desired position and velocity (as determined by the motion planner). Based on the errors between the actual and desired states, the controller determines what steering or velocity inputs are necessary to achieve the desired position and velocity.

At the lowest level are the actuator drivers. These controllers receive as input steering and velocity commands from the previous level controller and determine

P. Mellodge & P. Kachroo: Model Abstraction in Dynamical Systems, LNCIS 379, pp. 27–48, 2008.
springerlink.com

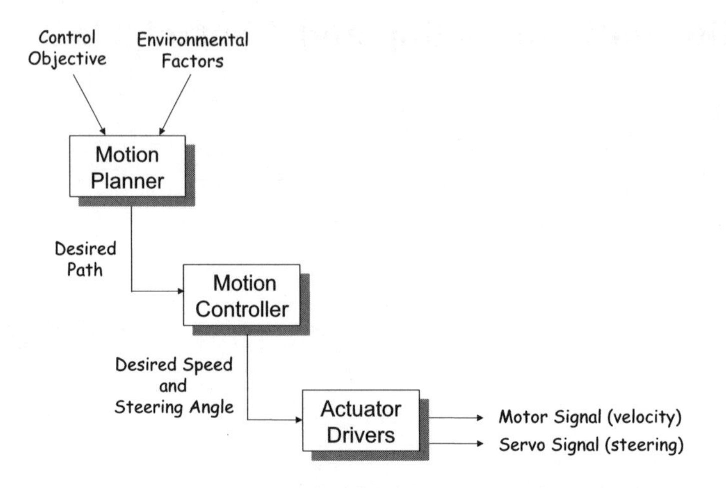

Fig. 3.1. The control hierarchy of a mobile robot

what inputs to the motors are necessary to achieve the desired position or rotational speed. In other words, the inputs are the desired steering angle and velocity from the motion controller. The actual steering angle and wheel speed are measured and compared with the desired values. This controller then determines the necessary motor and servo signals to achieve the desired values.

Because of this control structure, the different levels of controllers can be decoupled and designed separately. In this work, the design of the motion controller is studied. In particular, the path following problem is addressed. (The different types of motion controllers are described in Section 3.3.) Often, the mobile robot models used for the motion controller design are kinematic. Only the movement of the robot is modeled and the dynamic effects, such as mass or center of gravity, are not. The models used for the actuator drivers are dynamic as in [36]. The hierarchical structure allows for the robot dynamics to be compensated for at the lowest level.

3.2 Kinematic Modeling

There are two basic approaches to modeling a robotic car: kinematic and dynamic. A kinematic model assumes a no-slip condition on the wheels and considers only the movement of the vehicle. Acceleration is not accounted for in this model and it is assumed that changes in the velocity and steering angle are instantaneous. Under many conditions (such as on dry roads and when the car is being driven within its handling limits) these assumptions are valid and the kinematic model's behavior closely matches the actual car.

Dynamic modeling takes into account the forces acting on the vehicle. This model can be constructed using the no-slip condition [36] or allowing wheel slip [25]. In either case, the acceleration of the car is considered. In the case of [36], the design is used with the motion controller in the hierarchy described above and serves as the actuator controller. In [25], the model is used for traction control, where the objective is to steer the car on slippery surfaces or to achieve maximum acceleration (or deceleration).

The model used throughout this work is a kinematic model. This type of model allows for the decoupling of vehicle dynamics from its movement. Therefore, the vehicle's dynamic properties, such as mass, center of gravity, etc. do not enter into the equations. To derive this model, the nonholonomic constraints of the system are utilized.

3.2.1 Nonholonomic Contraints

As described in the previous chapter, if a system has restrictions in its velocity, but those restrictions do not cause restrictions in its positioning, the system is said to be nonholonomically constrained. Viewed another way, the system's local movement is restricted, but not its global movement. Mathematically, this means that the velocity constraints cannot be integrated to give position constraints.

A familiar example of a nonholonomic system is demonstrated by a parallel parking maneuver. When a driver arrives next to a parking space, he cannot simply slide his car sideways into the spot. The car is not capable of sliding sideways and this is the velocity restriction. However, by moving the car forwards and backwards and turning the wheels, the car can be placed in the parking space. Ignoring the restrictions caused by external objects, the car can be located at any position with any orientation, despite lack of sideways movement.

The nonholonomic constraints of each wheel of the mobile robot are shown in Figure 3.2. The wheel's velocity is in the direction of rolling. There is no velocity in the perpendicular direction. This model assumes that there is no wheel slippage.

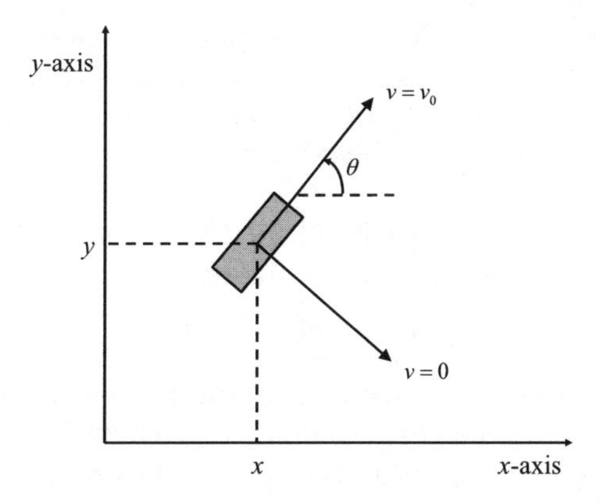

Fig. 3.2. The velocity constraints on a rolling wheel with no slippage

3.2.2 Frames of Reference

Before developing the models used for the mobile robot, it is appropriate to define the framework to be used throughout this work. In the following sections, two different frames of reference are used in describing the model: the global frame, F_g, and the mobile robot frame, F_m.

A top view the the mobile robot is given in Figure 3.3. The global frame, F_g, is fixed and the mobile robot frame, F_m, is attached to the robot and moves about in the global frame's (x, y) plane. The orientation of F_m is such that the linear velocity of the robot is along the x_m-axis. It is assumed that the robot's environment is such that there is no movement in the z-direction and that the z-axis and z_m-axis remain parallel at all times.

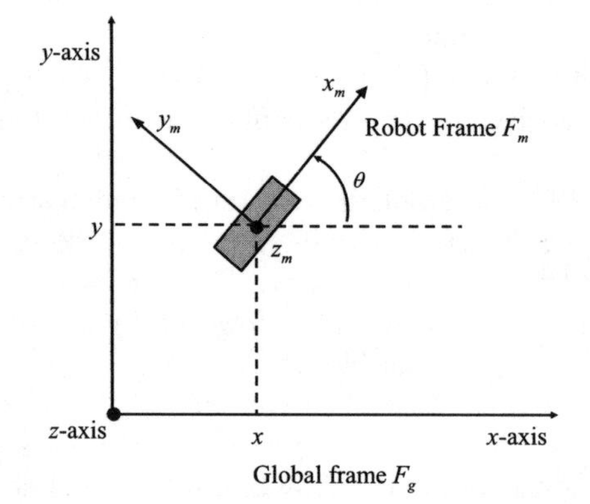

Fig. 3.3. Top view of the global and mobile robot frames

3.2.3 Unicycle Model

The simplest kinematic model for a mobile robot is given by the unicycle model, as shown in Figure 3.4. In this model, the z_m-axis goes through the point where the unicycle makes contact with the ground. Applying the nonholonomic constraints to the unicycle gives the following model:

$$\begin{bmatrix} \dot{x} \\ \dot{y} \\ \dot{\theta} \end{bmatrix} = \begin{bmatrix} v\cos\theta \\ v\sin\theta \\ \omega \end{bmatrix} \tag{3.1}$$

The inputs to the system are v and ω. The linear velocity, v, is in the direction of the x_m-axis and ω is the steering input that controls the angular velocity. In Section 2.2, it was shown using the Lie Algebra Rank Condition that the unicycle is controllable.

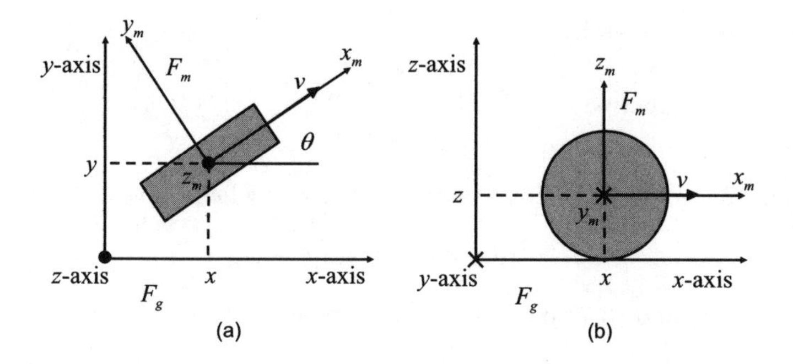

Fig. 3.4. The model for the unicycle. (a) Top view. (b) Side view.

3.2.4 Global Coordinate Model

The exact position and orientation of the car in the global coordinate system can be described by four variables, as shown in Figure 3.5. The (x, y) coordinates give the location of the center of the rear axle. The car's angle with respect to the x-axis is given by θ. The steering wheel's angle with respect to the car's longitudinal axis is given by ϕ.

From the constraints shown in Figure 3.2, the instantaneous velocity of the car in the x and y directions is given as

$$\dot{x} = v_1 \cos\theta \tag{3.2}$$

$$\dot{y} = v_1 \sin\theta \tag{3.3}$$

where v_1 is the linear velocity of the rear wheels.

The location of the center of the front axle (x_1, y_1) is given by

$$x_1 = x + l\cos\theta \tag{3.4}$$

$$y_1 = y + l\sin\theta \tag{3.5}$$

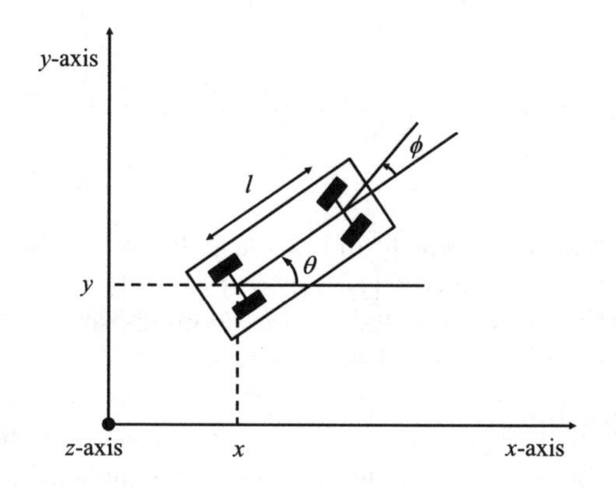

Fig. 3.5. The global coordinate system for the car

and the velocity is given by

$$\dot{x}_1 = \dot{x} - l\,\dot{\theta}\sin\theta. \tag{3.6}$$

$$\dot{y}_1 = \dot{y} + l\,\dot{\theta}\cos\theta \tag{3.7}$$

Applying the no-slippage constraint to the front wheels gives

$$\dot{y}_1 \cos(\theta+\phi) = \dot{x}_1 \sin(\theta+\phi). \tag{3.8}$$

Inserting (3.6) and (3.7) into (3.8) and solving for $\dot{\theta}$ yields

$$\dot{\theta} = \frac{\tan\phi}{l} v_1. \tag{3.9}$$

The complete kinematic model is then given as

$$\begin{bmatrix} \dot{x} \\ \dot{y} \\ \dot{\theta} \\ \dot{\phi} \end{bmatrix} = \begin{bmatrix} \cos\theta \\ \sin\theta \\ \frac{\tan\phi}{l} \\ 0 \end{bmatrix} v_1 + \begin{bmatrix} 0 \\ 0 \\ 0 \\ 1 \end{bmatrix} v_2 \tag{3.10}$$

where v_1 is the linear velocity of the rear wheels and v_2 is the angular velocity of the steering wheels.

Using the Lie Algebra Rank Condition, it can be shown that the robotic car is controllable. The vector fields f_1 and f_2 are given by

$$f_1 = \begin{bmatrix} \cos\theta \\ \sin\theta \\ \frac{\tan\phi}{l} \\ 0 \end{bmatrix} \quad \text{and} \quad f_2 = \begin{bmatrix} 0 \\ 0 \\ 0 \\ 1 \end{bmatrix}. \tag{3.11}$$

The Lie brackets $[f_1, f_2]$ and $[f_1, [f_1, f_2]]$ are given by

$$[f_1, f_2] = \begin{bmatrix} 0 \\ 0 \\ \frac{-1}{l\cos^2\phi} \\ 0 \end{bmatrix} \quad \text{and} \quad [f_1, [f_1, f_2]] = \begin{bmatrix} \frac{-\sin\theta}{l\cos^2\phi} \\ \frac{\cos\theta}{l\cos^2\phi} \\ 0 \\ 0 \end{bmatrix}. \tag{3.12}$$

The remaining Lie brackets are found to be $[0\;0\;0\;0]^T$. Upon computation we see that the rank of the matrix $[f_1, f_2, [f_1, f_2], [f_1, [f_1, f_2]]]$ is 4 except when $\phi = \pm\pi/2$. Thus the car is controllable away from those steering angles.

3.2.5 Path Coordinate Model

While the global model is useful for performing simulations, its use for path following is limited in practice. Often in the hardware implementation, the sensors cannot detect the car's location with respect to some global coordinates.

The sensors can only detect the car's location with respect to the desired path. Therefore, a more useful model is one that describes the car's behavior in terms of the path coordinates.

The path coordinates are shown in Figure 3.6. The perpendicular distance between the rear axle and the path is given by d. The angle between the car and the tangent to the path is $\theta_p = \theta - \theta_t$. The distance traveled along the path starting at some arbitrary initial position is given by s, the arc lengh.

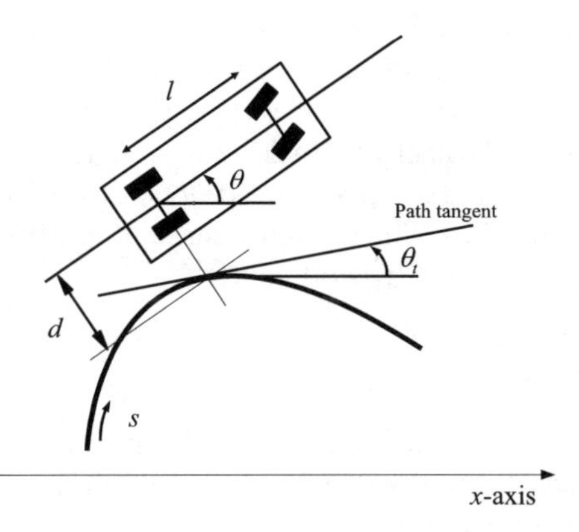

Fig. 3.6. The path coordinates for the car

According to [52], if it is assumed that the path to be followed is smooth and its curvature, denoted by $c(s)$, is differentiable, then the system can be transformed into the path coordinate model. The curvature is defined as

$$c(s) = \frac{d\theta_t}{ds}. \tag{3.13}$$

From this definition, $\dot{\theta}_t$ is given to be

$$\dot{\theta}_t = c(s)\dot{s}. \tag{3.14}$$

The velocity along the path is

$$\dot{s} = v_1 \cos\theta_p + \dot{\theta}_t d \tag{3.15}$$

and the velocity perpendicular to the path is

$$\dot{d} = v_1 \sin\theta_p. \tag{3.16}$$

Combining (3.14), (3.15), (3.16), and the definintion of θ_p, the car's kinematic model in terms of the path coordinates is given by [34]:

$$\begin{bmatrix} \dot{s} \\ \dot{d} \\ \dot{\theta}_p \\ \dot{\phi} \end{bmatrix} = \begin{bmatrix} \frac{\cos\theta_p}{1-dc(s)} \\ \sin\theta_p \\ \frac{\tan\phi}{l} - \frac{c(s)\cos\theta_p}{1-dc(s)} \\ 0 \end{bmatrix} v_1 + \begin{bmatrix} 0 \\ 0 \\ 0 \\ 1 \end{bmatrix} v_2 \tag{3.17}$$

where $c(s)$ is the path's curvature and is defined as

$$c(s) = \frac{d\theta_t}{ds}. \tag{3.18}$$

3.3 Control Design

There are three possible tasks that the car can perform: point-to-point stabilization, path following, and trajectory tracking. Point-to-point stabilization requires that the car move from point A to point B with no restrictions on its movement between those two points. With path following, the car must move along a geometric path. Trajectory tracking is similar to path following, except the car must follow a path at a given speed. The objective for the controller in this section is path following. The car must sense its position with respect to the path and return to the path if it is off course.

3.3.1 Chained Form

Before developing the controller for the model given in (3.17), the system is converted into chained form. As described in Section 2.4, the (2,n) single-chain form has the following structure:

$$\begin{aligned}
\dot{x}_1 &= u_1 \\
\dot{x}_2 &= u_2 \\
\dot{x}_3 &= x_2 u_1 \\
&\vdots \\
\dot{x}_n &= x_{n-1} u_1.
\end{aligned} \tag{3.19}$$

Although the system has two inputs, u_1 and u_2, this model can be considered single input if u_1 is known a priori.

For the car model with four states, the (2,4) chained form becomes

$$\begin{aligned}
\dot{x}_1 &= u_1 \\
\dot{x}_2 &= u_2 \\
\dot{x}_3 &= x_2 u_1 \\
\dot{x}_4 &= x_3 u_1.
\end{aligned} \tag{3.20}$$

The states are given as

$$\begin{aligned}
x_1 &= s \\
x_2 &= -c'(s) d \tan\theta_p - c(s)(1 - dc(s)) \frac{1 + \sin^2\theta_p}{\cos^2\theta_p} + \frac{(1 - dc(s))^2 \tan\phi}{l \cos^3\theta_p} \\
x_3 &= (1 - dc(s)) \tan\theta_p \\
x_4 &= d
\end{aligned} \tag{3.21}$$

where $c(s)$ is the path's curvature and $c'(s)$ denotes the derivative of c with respect to s.

The inputs are defined as follows:

$$\begin{aligned} v_1 &= \frac{1 - dc(s)}{\cos\theta_p} u_1 \\ v_2 &= \alpha_2(u_2 - \alpha_1 u_1) \end{aligned} \tag{3.22}$$

where v_1 is the linear velocity of the rear wheels, v_2 is the angular velocity of the steering wheels, and

$$\begin{aligned} \alpha_1 &= \frac{\partial x_2}{\partial s} + \frac{\partial x_2}{\partial d}(1 - dc(s))\tan\theta_p + \frac{\partial x_2}{\partial \theta_p}\left[\frac{\tan\phi(1 - dc(s))}{l\cos\theta_p} - c(s)\right] \\ \alpha_2 &= \frac{l\cos^3\theta_p\cos^2\phi}{(1 - dc(s))^2}. \end{aligned} \tag{3.23}$$

3.3.2 Derivation

With the system in chained form, a controller that performs path following can be developed. In this form, path following equates to stabilizing x_2, x_3, x_4 in (3.20) to zero. The input scaling controller from [34] is given here.

First the variables are redefined as follows:

$$\chi = (\chi_1, \chi_2, \chi_3, \chi_4) = (x_1, x_4, x_3, x_2). \tag{3.24}$$

So the chained form system is then

$$\begin{aligned} \dot{\chi}_1 &= u_1 \\ \begin{bmatrix} \dot{\chi}_2 \\ \dot{\chi}_3 \\ \dot{\chi}_4 \end{bmatrix} &= \begin{bmatrix} 0 & u_1 & 0 \\ 0 & 0 & u_1 \\ 0 & 0 & 0 \end{bmatrix} \begin{bmatrix} \chi_2 \\ \chi_3 \\ \chi_4 \end{bmatrix} + \begin{bmatrix} 0 \\ 0 \\ 1 \end{bmatrix} u_2. \end{aligned} \tag{3.25}$$

This system can be transformed into a linear time-invariant system if $u_1(t)$ is bounded and strictly positive (or negative) because such restrictions on u_1 would make χ_1 monotonically increasing (or decreasing) with time. So differentiation with respect to time can be replaced by

$$\frac{d}{dt}(\cdot) = \frac{d\chi_1}{dt}\frac{d}{d\chi_1}(\cdot) = u_1\frac{d}{d\chi_1}(\cdot). \tag{3.26}$$

Dividing by $|u_1|$ gives

$$\frac{1}{|u_1|}\frac{d}{dt}(\cdot) = \mathrm{sign}(u_1)\frac{d}{d\chi_1}(\cdot). \tag{3.27}$$

Then differentiating twice using this form of differentiation, the system can be represented by

$$\mathrm{sign}(u_1)\frac{d^3\chi_2}{d\chi_1^3} = \mathrm{sign}(u_1)\frac{u_2}{u_1} \tag{3.28}$$

and a stabilizing controller is given by

$$u_2(\chi_2, \chi_3, \chi_4) = -u_1 \text{sign}(u_1)[k_1\chi_2 + k_2 \text{sign}(u_1)\chi_3 + k_3\chi_4]. \tag{3.29}$$

As stated in [34], the system (3.25) is controllable if $u_1(t)$ is a "piecewise continuous, bounded, and strictly positive (or negative)" function. With u_1 known a priori, u_2 is left as the only input to the system. The controller for u_2 (with the appropriate restrictions on u_1) becomes

$$u_2 = -k_1|u_1(t)|\chi_2 - k_2u_1(t)\chi_3 - k_3|u_1(t)|\chi_4. \tag{3.30}$$

3.4 Curvature Estimation

The model for the car and the resulting controller given in the previous section require knowledge of the path's curvature. This section describes methods for estimating the path's curvature.

Except for $c(s)$, all of the variables in (3.17) are known or can be measured by sensors on the car. The feedback control algorithm based on this model must know the curvature to calculate the desired inputs v_1 and v_2. The problem then is to determine the curvature of the path based on the known or measured variables.

The task of curvature estimation is simplified if constraints are placed on the path. One such constraint is that the path be continuous. Another is that the path be either straight or a curve of known constant radius. A sample path showing these constraints is shown in Figure 3.7. This sample path is made up of straight sections and curves of two different radii. The resulting curvature profile is shown in Fig 3.8.

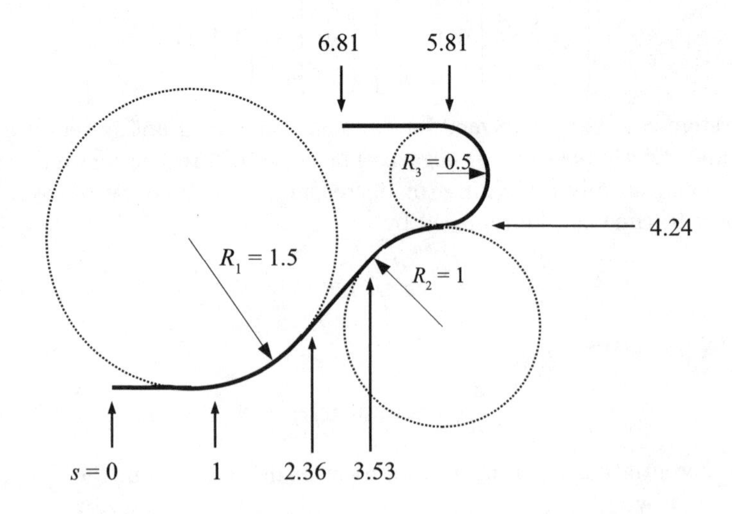

Fig. 3.7. A sample path showing the constraints

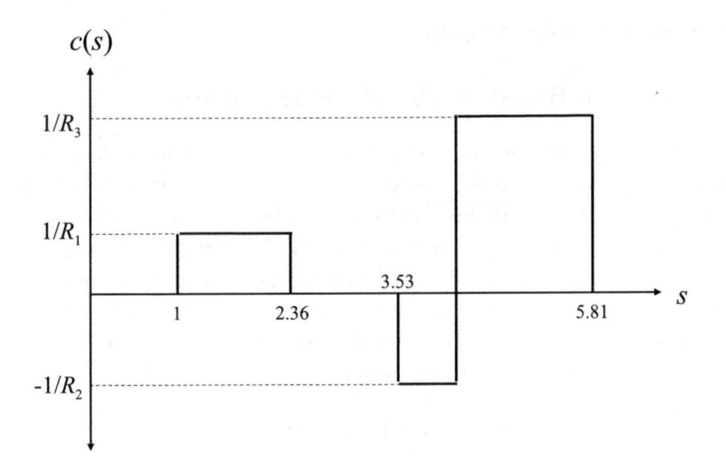

Fig. 3.8. The curvature of the path in Figure 3.7 with respect to the path length s

Recall from (3.18) that the curvature, $c(s)$, is defined as

$$c(s) = \frac{d\theta_t}{ds}.$$

Therefore, if the path is turning left, $c(s)$ is positive and if the path is turning right, $c(s)$ is negative. The magnitude of $c(s)$ is $\frac{1}{R}$, where R is the radius of the circle describing the curve.

As a result of these constraints, the curvature of the path as a function of distance is discontinuous and piecewise constant. The derivative of $c(s)$ with respect to distance is zero, except for those points where the curvature changes. At those points, the derivative is not defined. Therefore, the following assuption is made

$$\begin{aligned} c'(s) &= 0 \\ c''(s) &= 0 \\ &\vdots \end{aligned}$$

with $c'(s)$ denoting the derivative of c with respect to s. The derivatives are taken to be zero and points where the curvature changes are treated as disturbances.

If the curvature values are known a priori, the estimation result can be used to select the actual value. In other words, the calculated curvature need not be used for $c(s)$ in the state equations and controller. Rather, the actual curvature value can be selected based on the outcome of the estimation.

With the path configuration constraints defined, two methods of curvature estimation are now presented.

3.5 Estimation Methods

3.5.1 Estimation Based on the Steering Angle ϕ

The first method of estimating $c(s)$ is based solely on the steering angle, ϕ. At steady state, the car's steering wheels turn with the curves of the path. This method simply estimates the curvature using the steering wheels' angle.

If the front wheels are fixed at a certain angle, the car will describe a circle of a certain radius. Using (3.10), a MATLAB simulation was used to find the radius, R, described for several values of ϕ. It was found that the relationship between the circle's curvature, $c(s) = \frac{1}{R}$, and ϕ was nearly a straight line. So the relationship between $c(s)$ and ϕ was approximated to be

$$c(s) = \alpha + \beta\phi \tag{3.31}$$

where α and β were determined using the method of least squares to fit a line to the data. The sign of $c(s)$ is the same as ϕ.

To make this method more robust to noise, the value of ϕ used in (3.31) can be averaged over several sample periods. By averaging ϕ, this method provides a good estimate even if the car is oscillating about the desired path. However, (3.31) will work only if the car is generally following the desired path.

3.5.2 Estimation Based on the Vehicle Kinematics

The second method of estimating the curvature is based on the vehicle kinematics. If all the variables in (3.17) are known or can be measured, the equation can be solved for $c(s)$.

The third equation in (3.17) is

$$\dot{\theta}_p = \frac{v_1 \tan\phi}{l} - \frac{v_1 c(s) \cos\theta_p}{1 - dc(s)}. \tag{3.32}$$

This equation can be rearranged as

$$c(s)\left[v_1 \cos\theta_p + \frac{v_1 d \tan\phi}{l} - \dot{\theta}_p d\right] = \frac{v_1 \tan\phi}{l} - \dot{\theta}_p \tag{3.33}$$

which is linearly parameterizable in $c(s)$. This equation can be rewritten in the following form:

$$y = wa \tag{3.34}$$

where

$$y = \frac{v_1 \tan\phi}{l} - \dot{\theta}_p \tag{3.35}$$

$$w = v_1 \cos\theta_p + \frac{v_1 d \tan\phi}{l} - \dot{\theta}_p d \tag{3.36}$$

$$a = c(s). \tag{3.37}$$

Knowing w and y, a can be obtained using a least squares estimator. We want to find the $\hat{a}$ that minimizes J where

$$J = \int_0^t (y - w\hat{a})^2 dr. \tag{3.38}$$

Making $\frac{\partial J}{\partial \hat{a}} = 0$ gives

$$\left[\int_0^t w^2 dr\right] \hat{a} = \int_0^t wy dr. \tag{3.39}$$

Differentiating gives an update equation for $\hat{a}$:

$$\dot{\hat{a}} = -Pwe \tag{3.40}$$

where

$$P = \frac{1}{\int_0^t w^2 dr} \tag{3.41}$$

$$e = w\hat{a} - y \tag{3.42}$$

and w is defined in (3.36).

We can make the equation for P iterative by using the following update equation:

$$\dot{P} = -P^2 w^2 \tag{3.43}$$

where P is initialized to some large value.

3.6 Simulation Results

This section provides simulation results for the controller in (3.29) under varying conditions. The performance of the estimators is discussed and comparisons between them are made. The path used is shown in Figure 3.9, and its curvature profile is shown in Figure 3.10. This controller was tested using both the actual and the discretized errors.

3.6.1 Control Using the Actual Curvature

First, the actual curvature was used in the controller. The desired path is shown in Figure 3.9. The curvature can be determined to be 1 or 0. Figs. 3.11-3.13 show the results of applying this controller. The car's path speed, u_1, was held constant at 1.5 m/s. The gains used were $k_1 = \lambda^3$, $k_2 = 3\lambda^2$, and $k_3 = 3\lambda$, with $\lambda = 8$.

There are two important things to note about the performance of this controller. The first is that, even though u_1 is constant, v_1 does not remain constant. The control input u_1 is transformed into v_1 by taking into account the car's state and also the curvature. The result is that the car slows down in the curve.

The second thing to note is that there are spikes in the steering control input, u_2. These result from the spikes that are present in x_2. These spikes occur

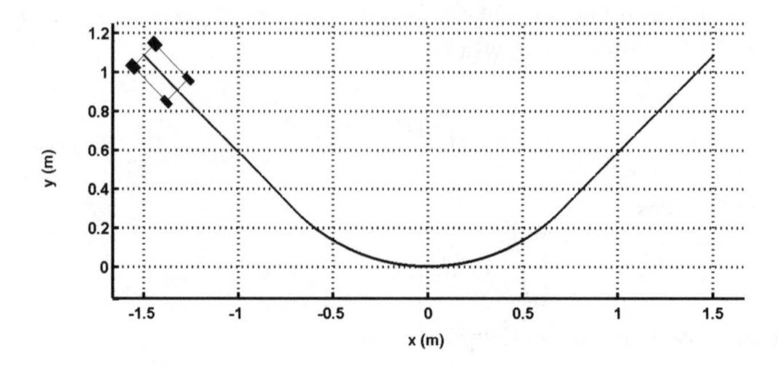

Fig. 3.9. The path used for simulation. Adapted from [24].

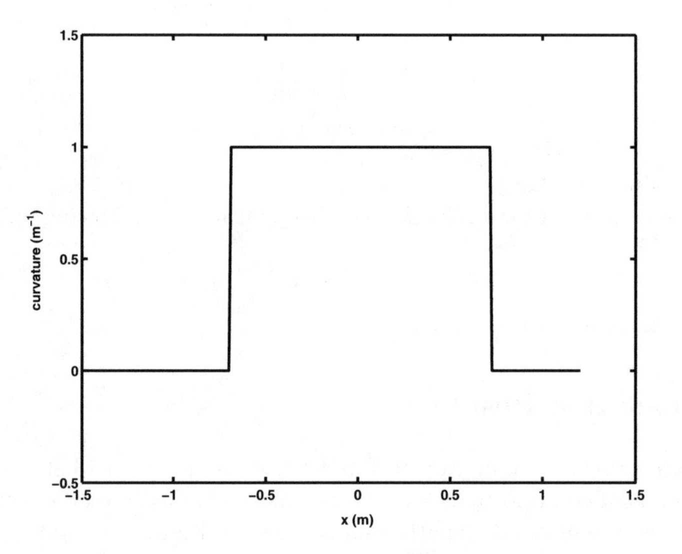

Fig. 3.10. The curvature profile of the path shown in Fig 3.9. Adapted from [24].

exactly where the path changes curvature. At these points, the curvature is not differentiable. However, in the implementation, the derivatives of curvature are set to zero. The discrepancy is seen here as a disturbance in the system.

Next, the same controller was used with the discretized errors. This simulates the errors obtained on the actual car using an array of discrete sensors. It was assumed that there were twelve sensors spaced 0.2 inches apart. The same gains and initial conditions were used as above. Figs. 3.14-3.16 show the results of discretization. Again, the actual curvature is used.

As is to be expected, using the discretized errors caused the control input, and thus the steering angle ϕ, to become much choppier. This resulted in a less smooth trajectory being traversed by the car.

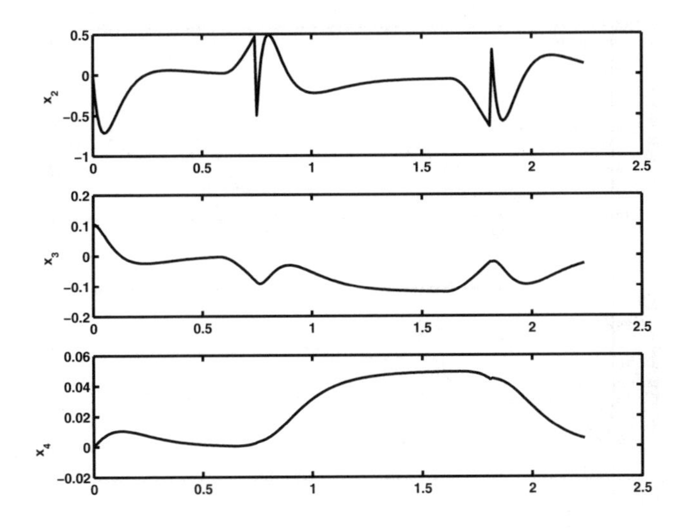

Fig. 3.11. The states, x_2, x_3, and x_4, resulting from using the actual errors and curvature. Adapted from [24].

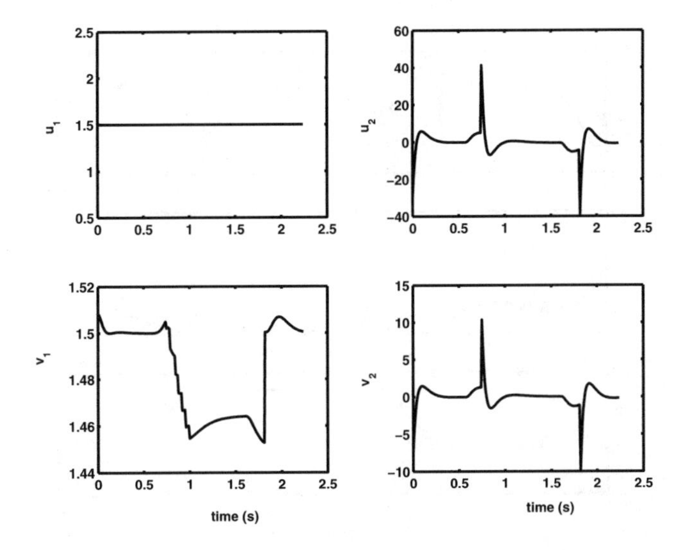

Fig. 3.12. The control inputs, v_1 and v_2, resulting from using the actual errors and curvature. Adapted from [24].

3.6.2 Control Using the ϕ Estimator

Next, the simulation was run using the ϕ estimator, as described in Section 3.5.1. The curvature was calculated using (3.31) with $\alpha = -0.1599$ and $\beta = 4.8975$. For filtering, ϕ was averaged over 10 sample periods. A threshold of 0.5 was used

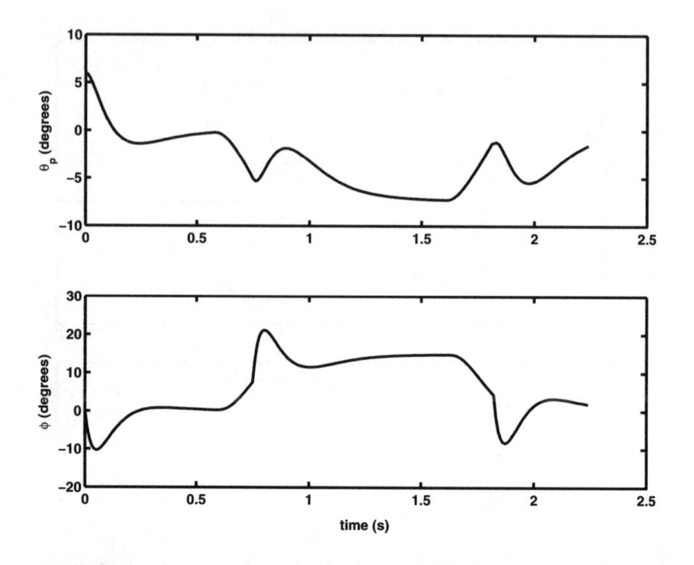

Fig. 3.13. The heading angle, θ_p, and steering angle, ϕ, resulting from using the actual errors and curvature. Adapted from [24].

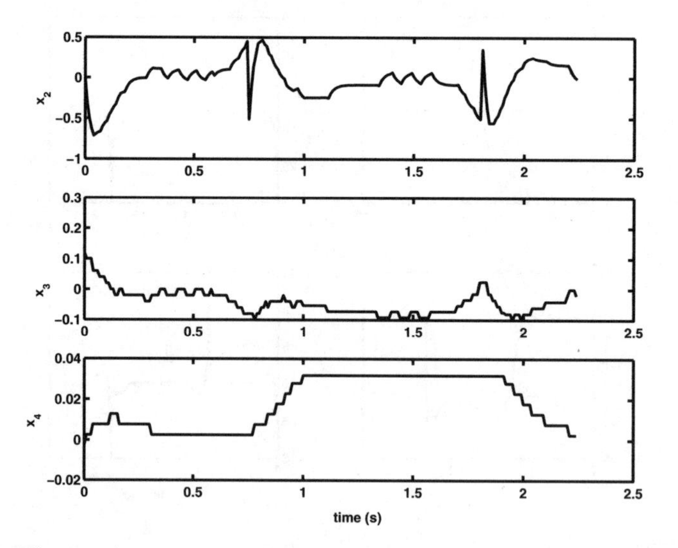

Fig. 3.14. The states, x_2, x_3, and x_4, resulting from using the discretized errors. Adapted from [24].

so that if the calculated curvature was less than 0.5, a $c(s)$ value of 0 was used. Otherwise a $c(s)$ value of 1 was used.

The car was placed on the path so that θ_p was initially nonzero. This initial condition resulted in some transients while the car centered itself on the path. The estimate of the curvature is shown in Figure 3.17a. The value used for $c(s)$ is

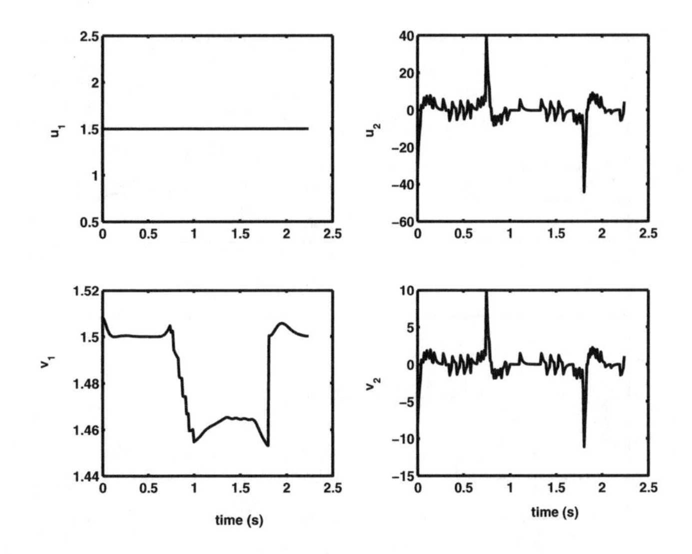

Fig. 3.15. The control inputs, v_1 and v_2, resulting from using the discretized errors. Adapted from [24].

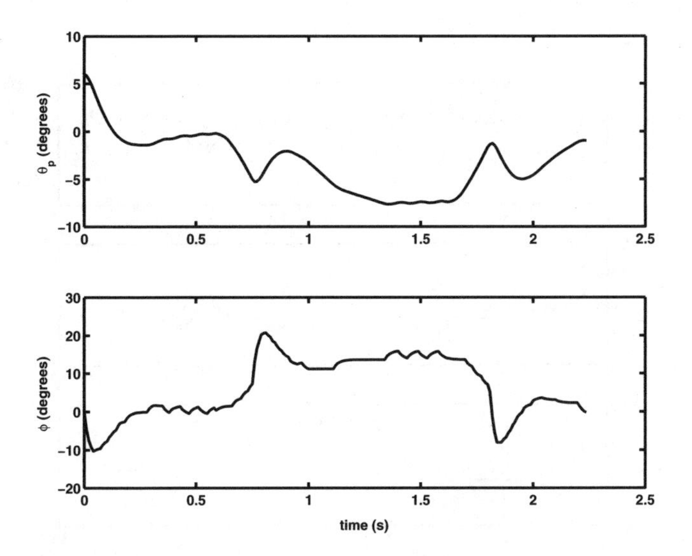

Fig. 3.16. The heading angle, θ_p, and the steering angle, ϕ, resulting from using the discretized errors. Adapted from [24].

shown as the solid line in Figure 3.17b. Because of the transients, this situation caused $c(s)$ to erroneously have a value of 1 well before the car reached the curve. This method gave a more accurate $c(s)$ during steady-state, showing only a slight delay. The results of control algorithm are shown in Figs. 3.18-3.20.

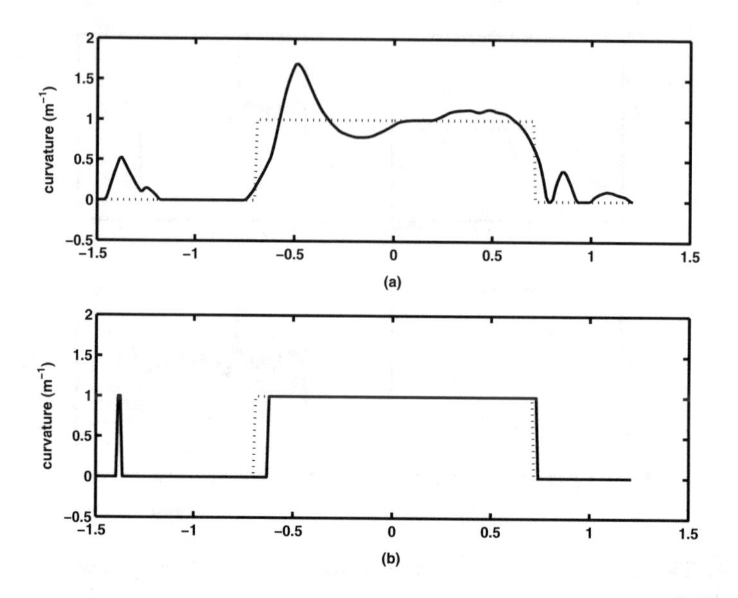

Fig. 3.17. The curvature estimated using only the steering angle, ϕ, with θ_p initially nonzero. Adapted from [24].

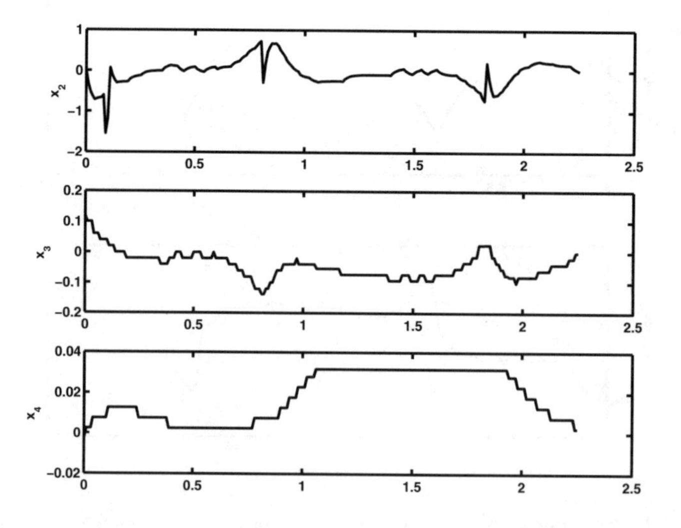

Fig. 3.18. The car's states resulting from the using the ϕ estimator. Adapted from [24].

In Figure 3.19, spikes are seen in the steering input on the straight portion of the path. Comparing this result with Figure 3.17 provides an explanation for this steering behavior. The ϕ estimator produced a false curvature while it was on the straightaway due to transients while the car corrected itself. This

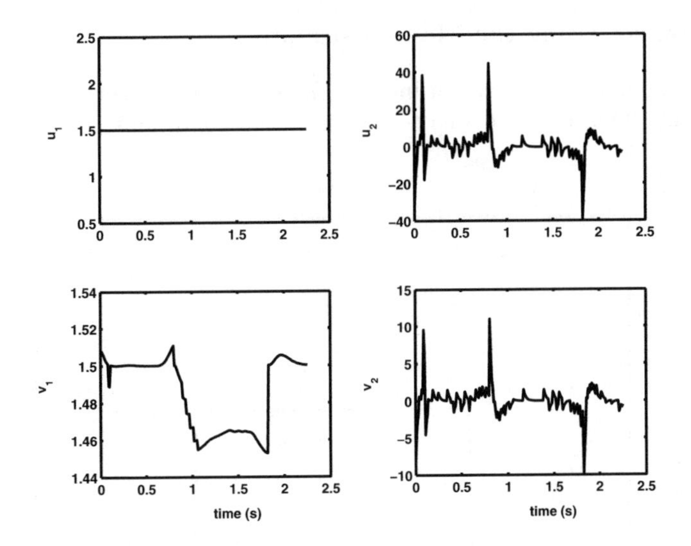

Fig. 3.19. The control inputs resulting from the using the ϕ estimator. Adapted from [24].

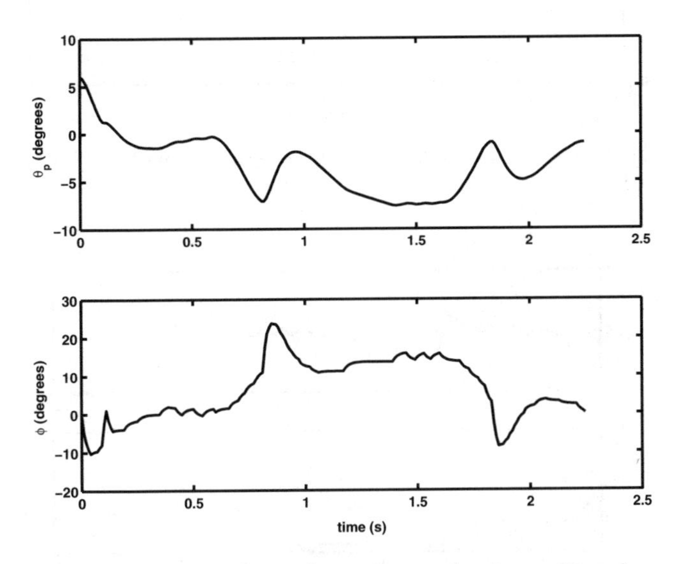

Fig. 3.20. The heading angle, θ_p, and steering angle, ϕ, resulting from using the ϕ estimator. Adapted from [24].

error in the curvature estimation results in two spikes in the steering input u_1, as seen in Figure 3.19. The two spikes occur in u_1 as it is starting out because the curvature is incorrectly estimated to be 1. The first spike occurs at the 0 to 1 transition, while the second occurs at the 1 to 0 transition. After the initial transients, this controller's performance was similar to the above controller with which the actual curvature was used.

3.6.3 Control Using the Model Estimator

Next, the simulation was run using the model estimate method described in Section 3.5.2. This method used the same initial conditions as the ϕ estimate method.

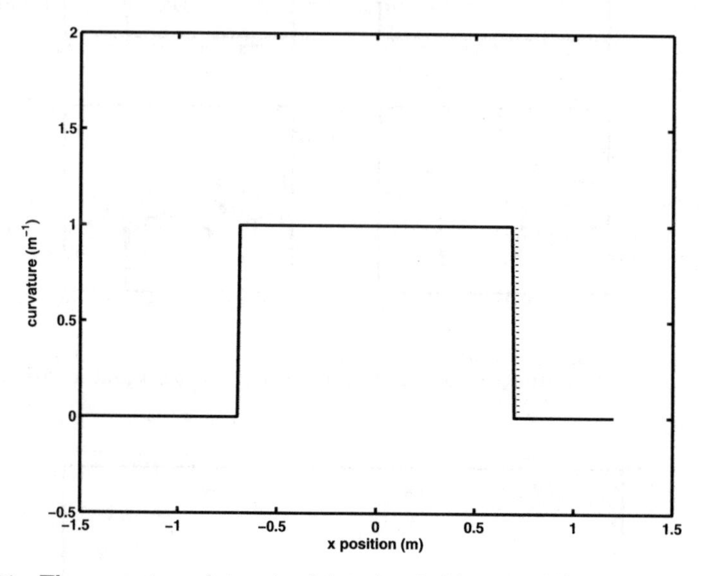

Fig. 3.21. The curvature determined by thresholding $\hat{a}$ with θ_p initially nonzero. Adapted from [24].

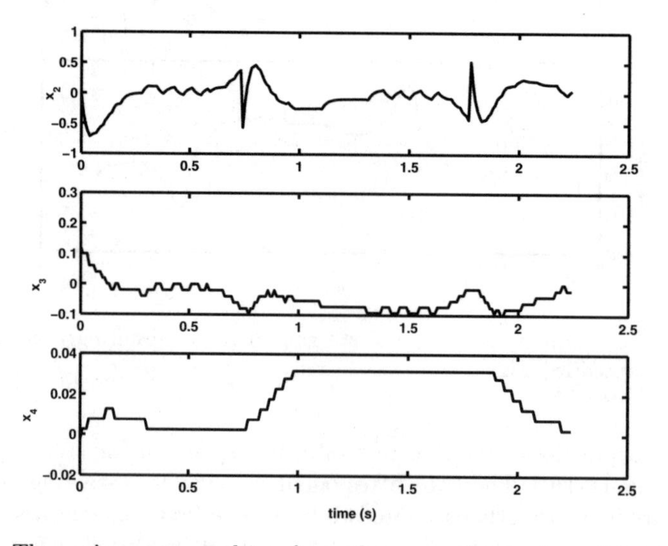

Fig. 3.22. The car's states resulting from the using the model estimator. Adapted from [24].

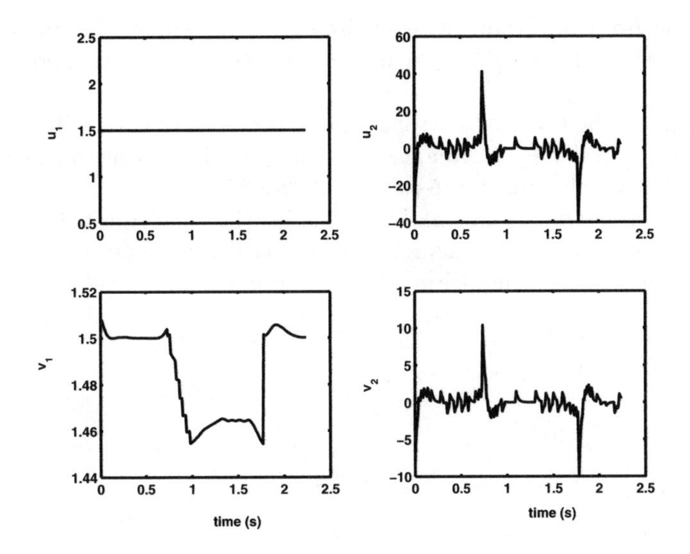

Fig. 3.23. The control resulting from the using the model estimator. Adapted from [24].

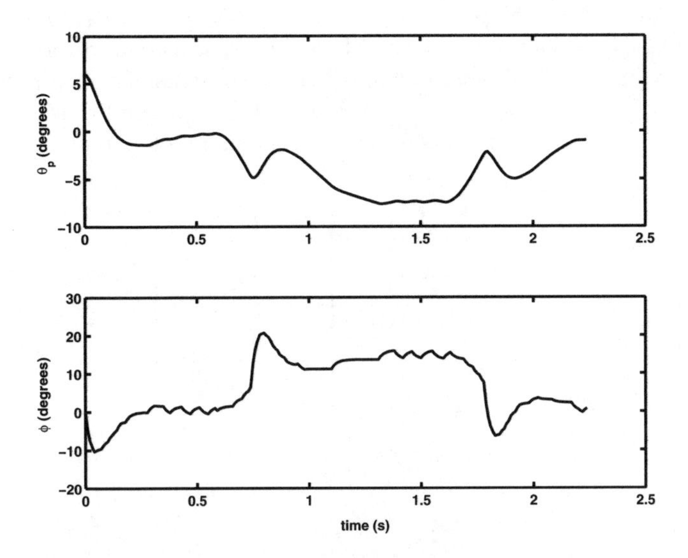

Fig. 3.24. The heading angle, θ_p, and steering angle, ϕ, resulting from using the model estimator. Adapted from [24].

As with the ϕ estimate method, this method was also tested with a nonzero θ_p. After applying the update equation, (3.40), $\hat{a}$ was thresholded. If it was greater than 0.5, it was set to 1. If it was less than 0.5, $\hat{a}$ was set to 0. The curvature value for $c(s)$ was then $\hat{a}$. The resulting curvature is shown in Figure 3.21. This method performed very well. The estimated curvature matched the actual curvature going from the straightaway to the curve. Coming out of the curve, there was only a slight delay before the estimator determined the correct value for $c(s)$.

As a result of the accurate curvature estimation, there is very little difference between the controller's performance using the actual curvature or the estimated curvature. The results are shown in Figs. 3.22-3.24.

This chapter presented kinematic models for the car-like robot: the unicycle model, the global coordinate model, and the path coordinate model. The next chapter presents another model in which the robot has a camera attached.

4 Vision Based Modeling and Control

4.1 Image Dynamic Modeling

Much of the work that has been done in the area of vision based control uses the image to obtain information about the car's location. For example, see [19], [20], and [58]. Often the car's position, angle, etc. are extracted from the image data. The idea of the model presented here (from [35]) is to bypass the extraction phase and use parameters directly measurable from the image to control the car.

In this chapter, it is assumed that there is a camera mounted rigidly on the car. The camera frame, F_c, is attached to the mobile robot as shown in Figure 4.1. This frame is chosen so that the origins of F_m and F_c are the same and that the y_m-axis lies along the y_c-axis. The camera is mounted at height h above the (x, y) plane. If, on the actual robot, the origins do not coincide, then a point in the actual camera frame can be transformed into F_c by a simple translation. Also, it is assumed that the camera is tilted downward so that α, the angle between the x_c-axis and x_m-axis, is positive. It is assumed that h and α are known fixed values. Throughout this chapter, points in the car's frame are denoted by (x_m, y_m, z_m) and points in the camera's frame are denoted by (x_c, y_c, z_c).

4.1.1 Ground Curves

In the previous chapter, to convert the kinematic model into path coordinate form it was assumed that the ground curve was smooth and its curvature differentiable. In this section, a precise statement of the conditions for the ground curve are given so that a transformation between the ground and image plane can be developed and the image dynamics studied.

The ground curve that the mobile robot is to follow will be denoted by Γ. If some assumptions are made about the ground curve, the analysis of these curves can be made easier. The assumptions are as follows:

1. Γ is analytic
2. Γ can be parameterized by z_c in the camera coordinates
3. $\frac{\partial \Gamma}{\partial z} = 0$

P. Mellodge & P. Kachroo: Model Abstraction in Dynamical Systems, LNCIS 379, pp. 49–59, 2008.
springerlink.com

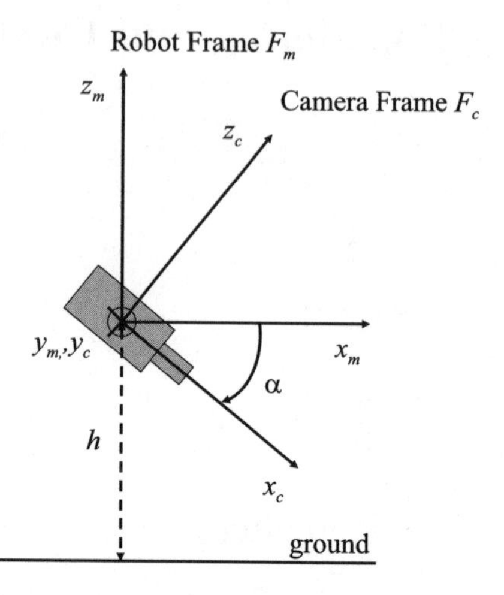

Fig. 4.1. The camera frame of reference and its relation to the robot frame

1. Γ is analytic
A curve is analytic at a point of Γ if it can be represented by a Taylor series with some radius of convergence. This implies the existence and continuity of all derivatives of Γ.

2. Γ can be parameterized by z_c
By this assumption, given any z_c, the corresponding y_c is unique. If the y_c coordinate were not unique, then there would be more than one possible path to follow. Thus a higher level decision would need to be made to determine in which direction to proceed.

3. $\frac{\partial \Gamma}{\partial z} = 0$
This is a formal statement of the previous assumption that the robot's environment is such that there is no movement in the z-direction and that the z-axis and z_m-axis remain parallel at all times.

The path coordinate model derived in the previous chapter is valid using these restrictions. This is because the transformation from global to path coordinates only require that the path be smooth and its curvature differentiable.

4.1.2 Image Dynamics

This section describes the relationship between the ground curve and the curve in the image plane and the dynamics of the image curve, as given in [35]. The coordinate notation has been modified to be consistent with the previous chapter.

By the second assumption made about the ground curve above, the points of Γ can be parameterized by z_c. Thus at any time t, a point on the ground curve, P, can be specified in the camera frame by $(\gamma_y(z_c, t), z_c, \gamma_x(z_c, t))$. An explicit expression can be derived for $\gamma_x(z_c, t)$. From Figure 4.2, it can be seen that

$$\gamma_x(z_c, t) = \frac{h}{\sin\alpha} + \frac{z_c}{\tan\alpha} = \frac{h + z_c\cos\alpha}{\sin\alpha}. \tag{4.1}$$

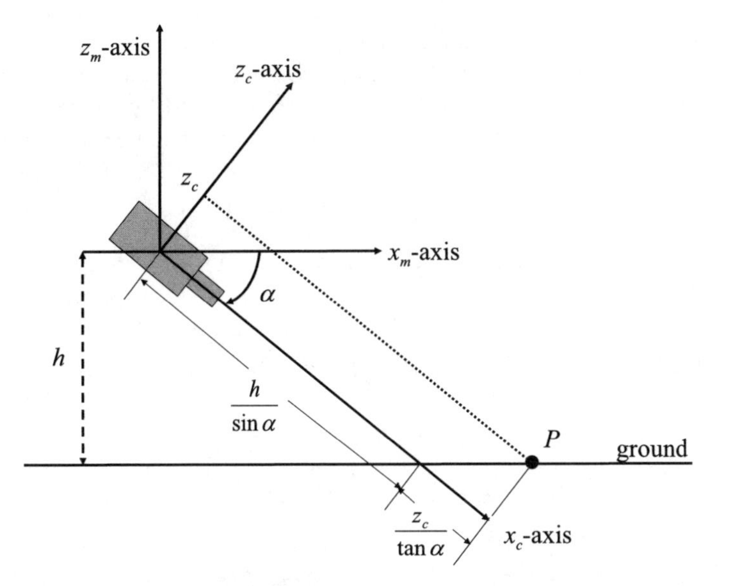

Fig. 4.2. Representation of point P on the ground curve in the camera coordinates

If z_c is a fixed point in the camera frame, the time evolution of the system can be characterized by $\gamma_y(z_c, t)$ since γ_x is not a function of time.

The ground curve can be represented in the image plane by its orthographic projection, $(\gamma_y(z_c, t), z_c)$, and also by its perspective projection, given by

$$Y(z_c, t) = f\frac{\gamma_y}{\gamma_x} \tag{4.2}$$

$$Z(z_c, t) = f\frac{z_c}{\gamma_x} \tag{4.3}$$

where f is the focal length of the camera. Inserting (4.1) into (4.2) and (4.3) gives

$$Y(z_c, t) = f\frac{\gamma_y(z_c, t)\sin\alpha}{h + z_c\cos\alpha} \tag{4.4}$$

$$Z(z_c, t) = f\frac{z_c\sin\alpha}{h + z_c\cos\alpha}. \tag{4.5}$$

At any time t, the derivative of Z with respect to z_c is given by

$$\frac{\partial Z(z_c, t)}{\partial z_c} = \frac{hf\sin\alpha}{(h + z_c\cos\alpha)^2}. \tag{4.6}$$

Since it is assumed that $\alpha \in (0, \frac{\pi}{2})$, if $z_c \neq -\frac{h}{\cos\alpha}$, then Z is a smooth function of z_c and $\frac{\partial Z(z_c,t)}{\partial z_c}$ is invertible. By the inverse function theorem [51], the mapping of z_c to Z is one-to-one in a neighborhood of P. Thus the image curve can be parameterized by Z and points of Γ can be represented in the perspective projection by $(\lambda_Y(Z, t), Z)$.

Since the ground curve is analytic and parameterized by z_c, γ_y is an analytic function of z_c and λ_Y is an analytic function of Z. Then both γ_y and λ_Y are infinitely differentiable and, at any z_c, can be uniquely determined by the values of their derivatives at z_c. Then the orthographic projection of Γ can be represented by

$$\begin{aligned} \xi_{i+1} &= \frac{\partial^i \gamma_y(z_c, t)}{\partial z_c^i}, \quad (i = 0, 1, ...) \\ \xi^i &\equiv (\xi_1, \xi_2, ..., \xi_i)^T \\ \xi &\equiv \xi^\infty. \end{aligned} \tag{4.7}$$

Similarly the perspective projection of Γ can be represented by

$$\begin{aligned} \zeta_{i+1} &= \frac{\partial^i \lambda_Y(Z, t)}{\partial Z^i}, \quad (i = 0, 1, ...) \\ \zeta^i &\equiv (\zeta_1, \zeta_2, ..., \zeta_i)^T \\ \zeta &\equiv \zeta^\infty. \end{aligned} \tag{4.8}$$

The two systems of equations, (4.7) and (4.8) are linearly related. If one set of coefficients is known, the other can be found through multiplication by a nonsingular lower triangular matrix. (The complete proof of this statement using mathematical induction is given in [35].) Thus, ζ_i is a function of $\xi_1, \xi_2, ..., \xi_i$ in general.

The 3x3 matrix relating ζ^3 and ξ^3 can be found as follows. By definintions (4.7) and (4.8),

$$\xi_1 = \gamma_y \tag{4.9}$$
$$\zeta_1 = \lambda_Y. \tag{4.10}$$

Plugging (4.4) into (4.10) gives

$$\zeta_1 = \frac{f \sin\alpha}{h + z_c \cos\alpha} \xi_1. \tag{4.11}$$

The relation between ξ_2 and ζ_2 can be found by differentiating ζ_1 with respect to z_c:

$$\zeta_2 \frac{\partial Z}{\partial z_c} = \frac{\partial \zeta_1}{\partial z_c}. \tag{4.12}$$

Substituting (4.11) into this equation and solving for ζ_2 gives the result in terms of ξ_1 and ξ_2

$$\zeta_2 = -\frac{\cos\alpha}{h}\xi_1 + \frac{h + z_c \cos\alpha}{h}\xi_2, \tag{4.13}$$

a linear combination of ξ_1 and ξ_2.

Similarly, ζ_3 is determined to be

$$\zeta_3 = \frac{(h + z_c \cos\alpha)^3}{f h^2 \sin\alpha}\xi_3. \tag{4.14}$$

Representing the above equations in matrix form gives the relationship between ξ^3 and ζ^3:

$$\zeta^3 = \begin{bmatrix} \frac{f \sin\alpha}{h+z_c \cos\alpha} & 0 & 0 \\ -\frac{\cos\alpha}{h} & \frac{h+z_c \cos\alpha}{h} & 0 \\ 0 & 0 & \frac{(h+z_c \cos\alpha)^3}{fh^2 \sin\alpha} \end{bmatrix} \xi^3. \tag{4.15}$$

In hardware implementation, the camera has a finite focal length so the measurements from the image will be the ζ coefficients, the perspective projection. However, the control laws and analysis to follow are done using the orthographic projection coefficients, ξ. Because of their equivalence from the above result, the properties of the orthographic projection system also hold for the perspective projection system.

The change in the arc length of a curve is given by [27]

$$s'(z_c) = \sqrt{\left(\frac{\partial \gamma_y}{\partial z_c}\right)^2 + 1 + \left(\frac{\partial \gamma_x}{\partial z_c}\right)^2} \tag{4.16}$$

The curvature of a ground curve was defined by (3.13) as

$$c(s) = \frac{d\theta_t}{ds} \tag{4.17}$$

This result only holds for planar curves. In the global frame, the Γ is planar. However in the camera frame, Γ moves in all three coordinates. For a curve in $\mathbb{R}^3$, the curvature (parameterized by z_c) is defined as

$$c(z_c) = \left|\frac{dT}{ds}\right| = \frac{|\Gamma'(z_c) \times \Gamma''(z_c)|}{|\Gamma'(z_c)|^3} \tag{4.18}$$

where T is the tangent vector of Γ at z_c. Since the derivatives of Γ are given by

$$\Gamma'(z_c) = \left(\frac{\partial \gamma_y}{\partial z_c}, 1, \cot\alpha\right)^T \tag{4.19}$$

$$\Gamma''(z_c) = \left(\frac{\partial^2 \gamma_y}{\partial z_c}, 0, 0\right)^T \tag{4.20}$$

then (4.18) becomes

$$c(z_c) = \frac{\sqrt{\cot^2\alpha + 1}\left(\frac{\partial^2 \gamma_y}{\partial z_c^2}\right)}{\left(\sqrt{\left(\frac{\partial \gamma_y}{\partial z_c}\right)^2 + 1 + \cot^2\alpha}\right)^3}. \tag{4.21}$$

Another important result from [35] concerns a special case of analytic ground curves: linear curvature curves. By definintion of linear curvature, the change in curvature with respect to the arc length is constant but not zero. Let $k = c'(s)$. Since $c'(s) = \frac{c'(z_c)}{s'(z_c)}$, rearranging and solving for $\frac{\partial^3 \gamma_y}{\partial z_c^3}$ results in

$$\frac{\partial^3 \gamma_y}{\partial z_c^3} = \xi_4 = \frac{\frac{k(1+\cot^2\alpha+\xi_2^2)^3}{\sqrt{1+\cot^2\alpha}} + 3\xi_2\xi_3^2}{1+\cot^2\alpha+\xi_2^2}. \tag{4.22}$$

This result shows that for $i \geq 4$, ξ_i is a function of only ξ_1, ξ_2, and ξ_3. So all of the information for a linear curvature analytic ground curve is captured in ξ^3.

As stated above, the dynamics of the image curve can be reduced to the dynamics of $\gamma_y(z_c, t)$ since γ_x is only a function of z_c. So as the mobile robot moves, $(\gamma_y(z_c, t), z_c, \gamma_x(z_c))$ changes with the rotational velocity ω and linear velocity v. The movement of a point in the camera frame due to the robot's linear velocity is given by

$$\begin{bmatrix} \dot{\gamma}_y \\ \dot{z}_c \\ \dot{\gamma}_x \end{bmatrix} = \begin{bmatrix} 0 \\ \sin\alpha \\ \cos\alpha \end{bmatrix} v. \tag{4.23}$$

The movement due to the robot's rotational velocity is given by [27]

$$\begin{bmatrix} \dot{\gamma}_y \\ \dot{z}_c \\ \dot{\gamma}_x \end{bmatrix} = - \begin{bmatrix} \gamma_y \\ z_c \\ \gamma_x \end{bmatrix} \times \begin{bmatrix} 0 \\ \omega\cos\alpha \\ -\omega\sin\alpha \end{bmatrix} \tag{4.24}$$

where $(0, \omega\cos\alpha, -\omega\sin\alpha)^T$ is the vector along the axis of rotation with length ω. Thus,

$$\dot{\gamma}_y = -(z_c \sin\alpha + \gamma_x \cos\alpha)\omega \tag{4.25}$$

$$\dot{z}_c = -(v\sin\alpha - \gamma_y \omega \sin\alpha). \tag{4.26}$$

Using the fact that,

$$\dot{\gamma}_y = \frac{\partial \gamma_y(z_c, t)}{\partial t} + \frac{\partial \gamma_y(z_c, t)}{\partial z_c} \dot{z}_c \tag{4.27}$$

the dynamics of a point in the image plane are

$$\frac{\partial \gamma_y(z_c, t)}{\partial t} = \left(\frac{\partial \gamma_y}{\partial z_c} \sin\alpha \right) v - \left(\frac{z_c}{\sin\alpha} + h\cot\alpha + \gamma_y \frac{\partial \gamma_y}{\partial z_c} \sin\alpha \right) \omega. \tag{4.28}$$

Thus $\dot{\xi}_1$ is given by (4.28). Differentiating twice with respect to z_c gives the image plane dynamics for a linear curvature curve:

$$\dot{\xi}^3 = f_1^3 \omega + f_2^3 v \tag{4.29}$$

where

$$f_1^3 = -\begin{bmatrix} \xi_1\xi_2 \sin\alpha + h\cot\alpha + \frac{z_c}{\sin\alpha} \\ \xi_1\xi_3 \sin\alpha + \xi_2^2 \sin\alpha + \frac{1}{\sin\alpha} \\ \xi_1\xi_4 \sin\alpha + 3\xi_2\xi_3 \sin\alpha \end{bmatrix} \quad (4.30)$$

$$f_2^3 = \begin{bmatrix} \xi_2 \sin\alpha \\ \xi_3 \sin\alpha \\ \xi_4 \sin\alpha \end{bmatrix}. \quad (4.31)$$

This system can be converted into chained form using the following transformations.

$$\begin{aligned}
x_1 &= \xi_2 \\
x_2 &= -\frac{\xi_3 \csc^3\alpha}{k(\csc^2\alpha + \xi_2^2)^3} \\
x_3 &= \xi_1 - \frac{\csc\alpha\, \xi_2\xi_3}{k(\csc^2\alpha + \xi_2^2)^2} \\
\omega &= \frac{-k\csc\alpha(\csc^2\alpha + \xi_2^2)^3 + 3\csc^2\alpha\, \xi_2\xi_3^2}{k(\csc^2\alpha + \xi_2^2)^4} u_1 - \frac{\xi_3}{\csc\alpha} u_2 \\
v &= \frac{-k\csc\alpha(\csc^2\alpha + \xi_2^2)^3}{k(\csc^2\alpha + \xi_2^2)^4} u_1 + \frac{3\csc^2\alpha\, \xi_2\xi_3(\csc^2\alpha + \xi_2^2 + \xi_3)}{k(\csc^2\alpha + \xi_2^2)^4} u_1 - \frac{\csc^2\alpha + \xi_2^2 + \xi_3}{\csc\alpha} u_2
\end{aligned} \quad (4.32)$$

4.2 Unicycle Controller

For the system given by (4.29), if the robot is tracking the desired curve, then the lateral deviation of the wheel from the curve should be zero and the robot should be moving in the same direction as the ground curve's tangent. Mathematically this is expressed as

$$\begin{aligned}
\gamma_y(z_c, t)|_{z_c = -h\cos\alpha} &= 0 \\
\frac{\partial \gamma_y(z_c, t)}{\partial z_c}|_{z_c = -h\cos\alpha} &= 0.
\end{aligned} \quad (4.33)$$

Thus the controller should stabilize ξ_1 and ξ_2 to zero.

Combining (4.33) with (4.29) and solving for ω gives the angular velocity when the mobile robot is tracking the desired curve:

$$\omega = (\xi_3 \sin^2\alpha) v. \quad (4.34)$$

The control law from [35] is given by

$$\begin{aligned}
\omega &= \xi_3 \sin^2\alpha\, v_0 + \sin^2\alpha\, \xi_1 v_0 + K_\omega \xi_2 \\
v &= v_0 + \sin^2\alpha\, \xi_1(\xi_1 + \xi_3) v_0 - K_v \xi_2 \mathrm{sign}(\xi_1 + \xi_3).
\end{aligned} \quad (4.35)$$

If K_ω and K_v are positive, (4.35) is a stable controller for system (4.29) about $\xi_1 = \xi_2 = 0$ given the desired linear velocity v_0. This can be shown using the Lyapunov stability theorem on the partial system $(\xi_1, \xi_2)^T$ with the Lyapunov function defined as

$$V = \xi_1^2 + \xi_2^2. \tag{4.36}$$

Then differentiation gives

$$\dot{V} = -\frac{K_\omega \xi_2^2}{\sin \alpha} - K_\omega \sin \alpha \xi_2^4 - \xi_2^2 \sin \alpha [(K_\omega \xi_1 + \xi_2 \sin^2 \alpha v_o) + K_v \mathrm{sign}(\xi_1 + \xi_3)](\xi_1 + \xi_3). \tag{4.37}$$

If ξ_1 and ξ_2 are small enough, then

$$|K_\omega \xi_1 + \xi_2 \sin^2 \alpha \, v_0| \leq K_v \tag{4.38}$$

making

$$\dot{V} \leq -\frac{K_\omega \xi_2^2}{\sin \alpha} \leq 0. \tag{4.39}$$

So by Lyapunov's stability theorem, (4.29) with the controller given by (4.35) is stable about $\xi_1 = \xi_2 = 0$.

4.3 Simulation Results

This section provides simulation results for the controller in (4.35) for the case where the ground curve has constant curvature ($c'(s) = 0$) and linear curvature ($c'(s)$ is a nonzero constant). The ground curves to be followed by the unicycle are shown in Figs. 4.3 and 4.4. For the circular path, the radius of curvature was 3. For the spiral path, the radius of curvature was given by $\frac{3}{s}$, where $s \in [1, 2]$.

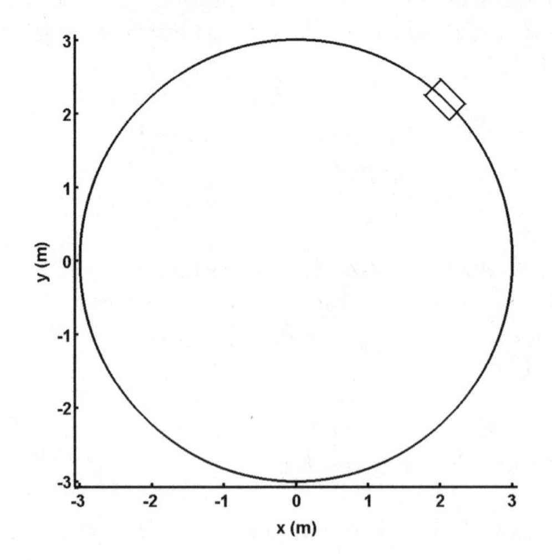

Fig. 4.3. The circular path to be followed by the unicycle

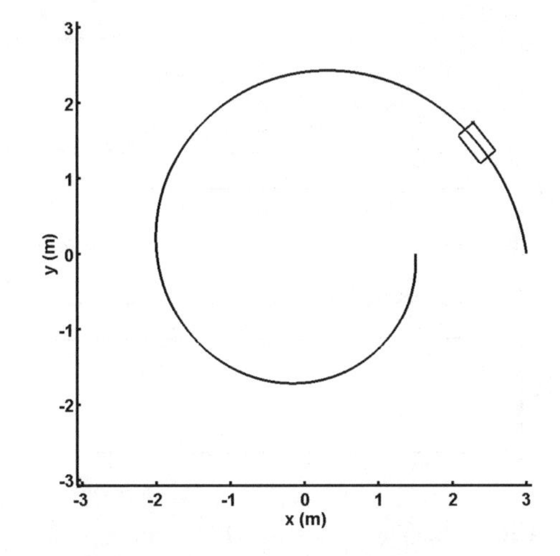

Fig. 4.4. The spiral path to be followed by the unicycle

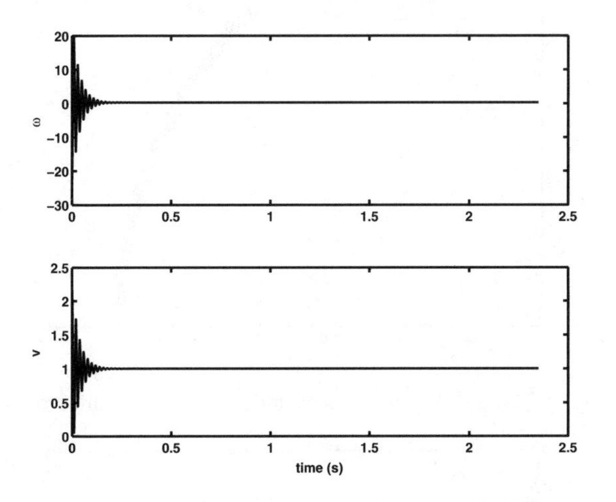

Fig. 4.5. The inputs for the circular path

In both cases, the values $K_\omega = 200$, $K_v = 10$, and $\alpha = \frac{\pi}{3}$ were used and the initial position was given by $x_0 = 3$, $y_0 = 0.1$, and $\theta_0 = 100°$.

The results of applying the controller to the unicycle on the circular path are shown in Figs. 4.5, 4.6, and 4.7. As can be seen in Figure 4.6, the controller drives ξ_1 and ξ_2 to zero, which indicates tracking. The value of ξ_3 settles at 0.5, which results from the unicycle being aligned with the path's tangent.

The results of applying the controller to the unicycle on the spiral path are shown in Figs. 4.8, 4.9, and 4.10. As with the circular path, the controller drives ξ_1 and ξ_2 to zero, indicating that the unicycle is tracking the path. The value of ξ_3, however, is increasing linearly. This indicates that the unicycle is aligned with the path, whose curvature in increasing linearly.

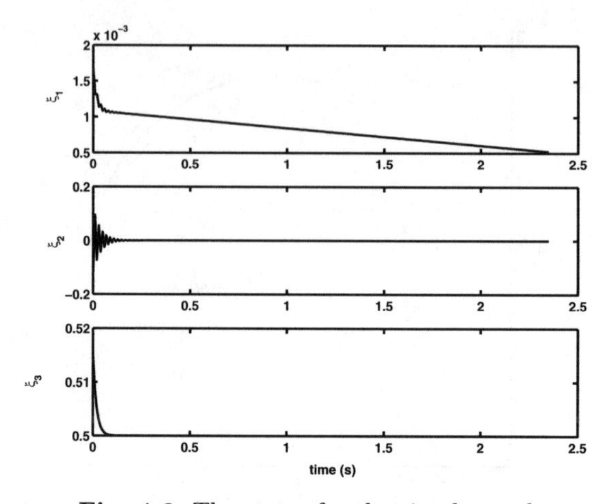

Fig. 4.6. The states for the circular path

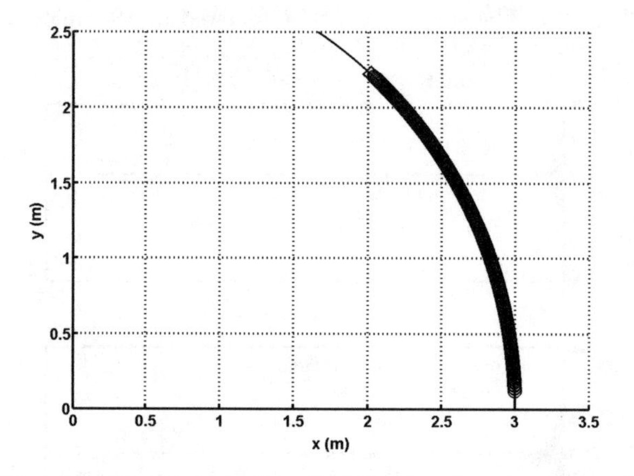

Fig. 4.7. The unicycle's path following the circular path

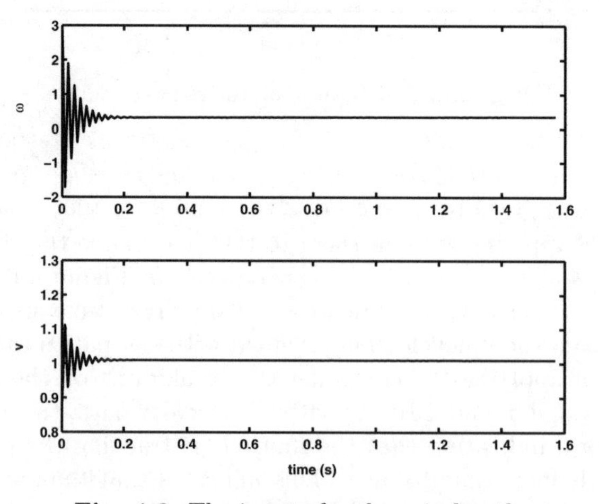

Fig. 4.8. The inputs for the spiral path

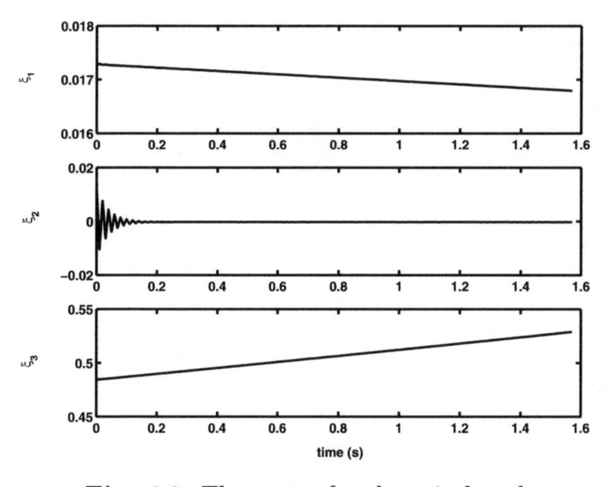

Fig. 4.9. The states for the spiral path

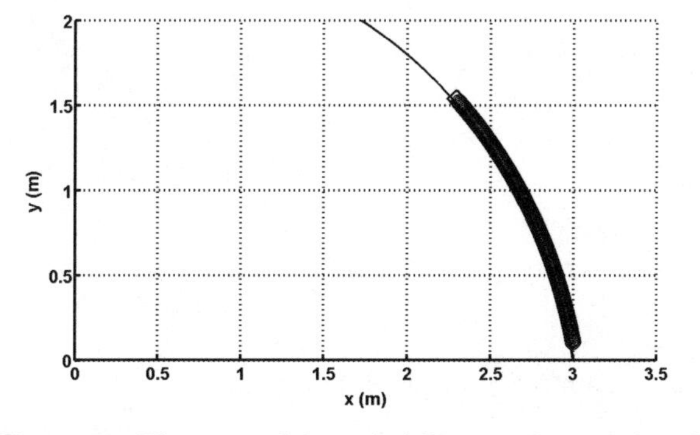

Fig. 4.10. The unicycle's path following the spiral path

The controller design given in this chapter is for the unicycle. However, the goal is to design a controller for the car. The next chapter investigates the relationship between the unicycle and the car in terms of abstraction, a concept that allows dynamical systems such as the car to be represented by simpler systems such as the unicycle.

5 Abstraction

This chapter reviews the concept of abstracted control systems and introduces the new concepts of traceability, ϵ-traceability, and ϵ-consistency. Abstracted systems are simplifications of more complex dynamical systems that retain some important information, such as controllability, about the original system. The notions of traceability, ϵ-traceability, ϵ-consistency deal with the relationship between two dynamical systems. This chapter discusses these concepts and applies them to the robotic car and unicycle systems. It starts with a review of several concepts used in the definition of abstraction. A full discussion of these concepts can be found in [47].

5.1 Preliminary Definitions

Abstracted control systems deals with relationships between systems. It is necessary to have a transformation from one system to the other that is well-defined. The following definition from [1] ensures that for any vector field X, the tangent map $T\Phi(X(p))$ is well-defined for all $p \in M$.

Definition 5.1. ***Φ-Related Vector Fields:*** *Let $\Phi : M \to N$ be a smooth map and let X and Y be vector fields on M and N, respectively. Then X and Y are Φ-related if for all $p \in M$*

$$T\Phi(X(p)) = Y(\Phi(p)). \tag{5.1}$$

If $\Phi : M \to N$ is a smooth surjection[1], then for any vector field X, $T\Phi(X)$ is well-defined on N only if

$$T_{p_1}\Phi(X(p_1)) = T_{p_2}\Phi(X(p_2)) \tag{5.2}$$

whenever $\Phi(p_1) = \Phi(p_2)$ for any $p_1, p_2 \in M$. Figure 5.1 shows the relationship between vector fields in manifolds M and N. The vector fields X and Y are Φ-related if $T\Phi(X(p)) = Y(\Phi(p))$ for all $p \in M$.

[1] A mapping $\Phi : M \to N$ is surjective if for every $q \in N$ there exists a point $p \in M$ such that $\Phi(p) = q$.

P. Mellodge & P. Kachroo: Model Abstraction in Dynamical Systems, LNCIS 379, pp. 61–80, 2008.
springerlink.com

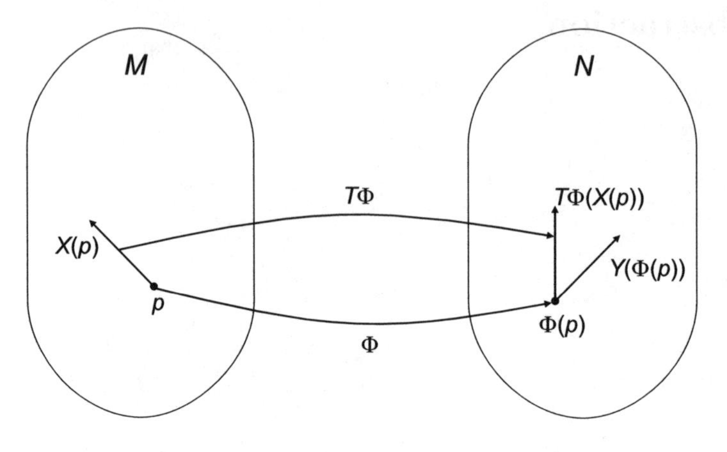

Fig. 5.1. The relationship between vector fields in M and N. The vector fields X and Y are Φ-related if $T\Phi(X(p)) = Y(\Phi(p))$.

It is useful to view Φ-relatedness of vector fields in terms of their integral curves. The condition on the integral curves of two Φ-related vector fields is given by the following theorem from [1].

Theorem 5.2. *Suppose X and Y are vector fields on M and N respectively and let $\Phi : M \to N$ be a smooth map, then X and Y are Φ-related if and only if for every integral curve c on M, $\Phi \circ c$ is an integral curve on N.*

Next the concept of Φ-relatedness of vector fields is extended to control systems. In order to have a coordinate free and general description of control systems, the concept of fiber bundles is used. Let us consider nonlinear control systems described by the ordinary differential equations

$$\dot{x}(t) = f(x(t), u(t)) \tag{5.3}$$

where $x(t) \in \mathbb{R}^n$ denotes the state of the system and $u(t) \in \mathbb{R}^m$ denotes the control input. The control system in (5.3) can be described using the commutative diagram in Figure 5.2 [44]. In the figure,

$$\begin{aligned} (i_d, f)(x(t), u(t)) &= (x(t), \dot{x}(t)) \\ \pi(x, u) &= x \\ \pi'(x) &= (x, \dot{x}). \end{aligned} \tag{5.4}$$

This control system representation can be generalized from the above local description to a global one on a manifold M as shown in Figure 5.3 [44]. The components of the control system can be described as follows.

M	the state space manifold
$\pi : B \to M$	a fiber bundle
$\pi^{-1}(x)$	the state dependent input spaces where $x \in M$

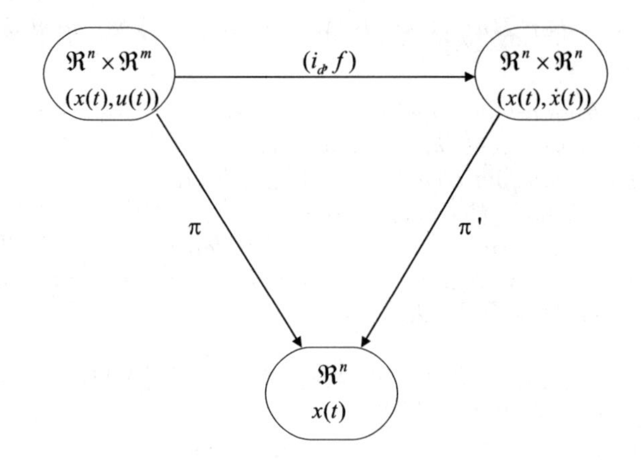

Fig. 5.2. Local description of a control system. Adapted from [44].

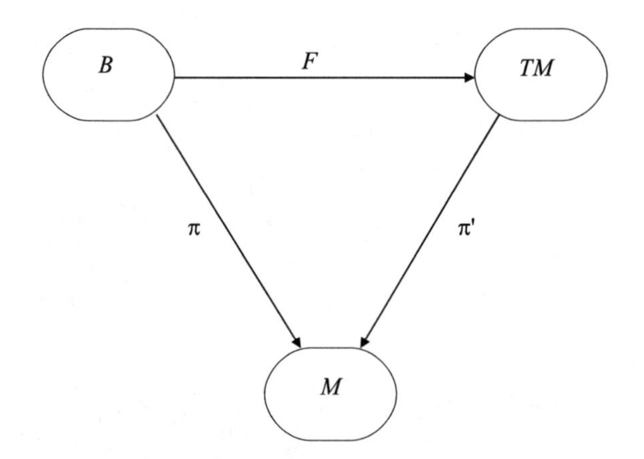

Fig. 5.3. Global description of a control system. Adapted from [44].

TM the tangent space of M
$\pi' : TM \to M$ the canonical projection of TM on M
$F : B \to TM$ the system dynamics

In other words, for any point (x, u) in B, $F(x, u) = (x, f(x, u))$ where $f(x, u)$ is the velocity vector at the point $x \in M$. Using the local coordinates of the manifolds, this representation is exactly the same as given above.

The fiber bundle $\pi : B \to M$ is often simply $M \times U$ for some input space U. However, this is generally not the case because the input space may be dependent on the state. The use of product $M \times U$ is then insufficient to describe the space. The product space can be generalized into the form of a fiber bundle, defined as follows.

Definition 5.3. ***Fiber Bundles:*** *$(B, M, \pi, U, \{O_i\}_{i \in I})$ is called a fiber bundle if the following holds.*

i. B is a smooth manifold called the total space.
ii. M is a smooth manifold called the base space.
iii. U is a smooth manifold called the standard fiber.
iv. $\pi : B \to M$ is a surjective submersion[2].
v. $\{O_i\}_{i \in I}$ is an open cover for M such that for every $i \in I$, there exists a diffeomorphism $\Psi_i : \pi^{-1}(O_i) \to O \times U$ such that $\pi_o \circ \Psi_i = \pi$, where π_o is the projection from $O_i \times U$ to O_i.

The submanifold $\pi^{-1}(p)$ is called the fiber at $p \in M$. Figure 5.4 illustrates the relationships of a fiber bundle [44].

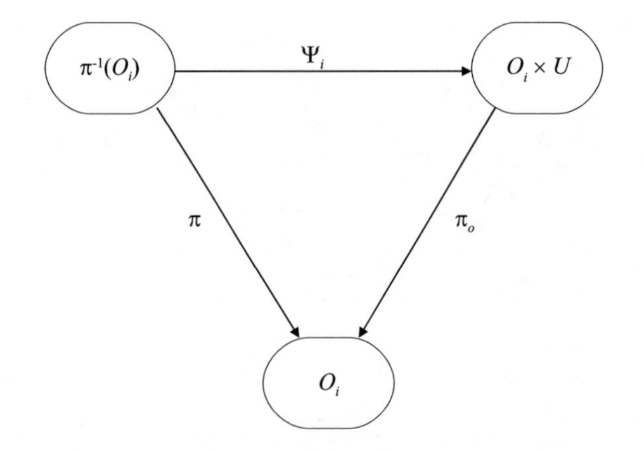

Fig. 5.4. Fiber bundle. Adapted from [44].

Definition 5.4. ***Control Systems:*** *A control system $S = (B, F)$ consists of the following.*

1. A fiber bundle $\pi : B \to M$ called the control bundle.
2. A smooth map $F : B \to TM$ which is fiber preserving.

A map is fiber preserving if $\pi' \circ F = \pi$, where $\pi' : TM \to M$ is the tangent bundle projection. Figure 5.3 illustrates globally the relationships of a control system and Figure 5.2 shows the local relationships for an n-dimensional system with m inputs.

Definition 5.5. ***Trajectories of Control Systems:*** *Let $I \subseteq \mathbb{R}$. A smooth curve $c : I \to M$ is called the trajectory of the control system (B, F) if there exists a curve $c^B : I \to B$ such that $\pi(c^B) = c$ and $c'(t) = F(c^B(t))$.*

[2] A smooth map $\pi : B \to M$ is a submersion if the rank of the Jacobian of π at every point in B is equal to the dimension of M.

In local coordinates, this definition translates into the well-known concept of trajectories in a control system given by $\dot{x} = f(x, u, t)$. Locally, a trajectory is a curve $x(t)$ for which there exists an input $u(t)$ such that $\dot{x}(t) = (x(t), u(t), t)$.

Definition 5.6. ***Φ-Related Control Systems:*** *Let $\Phi : M \to N$ be a smooth map and let $S_M = (B_M, F_M)$ and $S_N = (B_N, F_N)$ be two control systems with $\pi_M : B_M \to M$ and $\pi_N : B_N \to N$. Then control systems S_M and S_N are Φ-related if*

$$T\Phi \circ F_M(\pi_M^{-1}(p)) \subseteq F_N(\pi_N^{-1}(\Phi(p))) \tag{5.5}$$

for all $p \in M$.

It follows from the definition of Φ-related control systems that any trajectory in S_M, when mapped to N pointwise through Φ, gives a trajectory that can be obtained by a control input in S_N. This extension to trajectories is formalized in the following theorem from [47].

Theorem 5.7. ***Trajectories of Φ-Related Control Systems:*** *Let $\Phi : M \to N$ be a smooth map and let $S_M = (B_M, F_M)$ and $S_N = (B_N, F_N)$ be two control systems. Then S_M and S_N are Φ-related if and only if for every trajectory $c_M : I \to M$, $\Phi \circ c_M$ is a trajectory of S_N.*

The following example, motivated by the double integrator example in [47], illustrates concept of Φ-related control systems in higher dimensions.

Example 5.8. Consider the control system, S_M, given by

$$\begin{bmatrix} \dot{x}_1 \\ \dot{x}_2 \\ \dot{x}_3 \end{bmatrix} = \begin{bmatrix} x_2 \\ x_3 \\ u \end{bmatrix} \tag{5.6}$$

where $M = \mathbb{R}^3$ and $u \in \mathbb{R}$. Let Φ be the projection map defined by $\Phi(x_1, x_2, x_3) = (x_1, x_2)$. Define the control system, S_N, to be

$$\begin{bmatrix} \dot{x}_1 \\ \dot{x}_2 \end{bmatrix} = \begin{bmatrix} x_2 \\ v \end{bmatrix} \tag{5.7}$$

where $N = \mathbb{R}^2$ and $v \in \mathbb{R}$. Let $p = (x_1, x_2, x_3)$ be a point in M. Then vector fields in S_N mapped from vector fields in S_M are found by the following process.

$$\begin{gathered} p = (x_1, x_2, x_3) \in M \\ \Downarrow \pi_M^{-1} \\ (x_1, x_2, x_3, u) \in B_M \\ \Downarrow F_M \\ \left(x_1, x_2, x_3, [x_2, x_3, u]^T\right) \in TM \end{gathered}$$

$$\Downarrow T\Phi$$

$$\left(x_1, x_2, [x_2, x_3]^T\right) \in TN$$

Similarly, vector fields in S_N originating from $\Phi(p)$ are found as follows.

$$p = (x_1, x_2, x_3) \in M$$

$$\Downarrow \Phi$$

$$(x_1, x_2) \in N$$

$$\Downarrow \pi_N^{-1}$$

$$(x_1, x_2, v) \in B_M$$

$$\Downarrow F_N$$

$$\left(x_1, x_2, [x_2, v]^T\right) \in TN$$

Setting $v = x_3$ generates the corresponding vector field in N. Because of the freedom in the control input v, S_N can generate any vector field that is mapped from S_M as illustrated in Figure 5.5. So it follows that $T\Phi \circ F_M(\pi_M^{-1}(p)) \subseteq F_N(\pi_N^{-1}(\Phi(p)))$ and control systems S_M and S_N are Φ-related because $v \in \mathbb{R}$.

In addition to constructing abstractions it is also desirable to propagate properties between systems. In particular the property of controllability is reviewed here. The following definitions of reachability are utilized towards that end.

Definition 5.9. ***Reachability:*** *Let $S = (B, F)$ be a control system on a manifold M. Given a point $p \in M$, the reachable set in time $t \in I \subseteq \mathbb{R}$ is defined as*

$$\begin{aligned}\text{Reach}(p, S) = \{p' \in M \mid & \text{ there exists } c : I \to M \text{ such that} \\ & c(t_1) = p \text{ and } c(t_2) = p', \text{ for } t_1, t_2 \in I\}.\end{aligned} \tag{5.8}$$

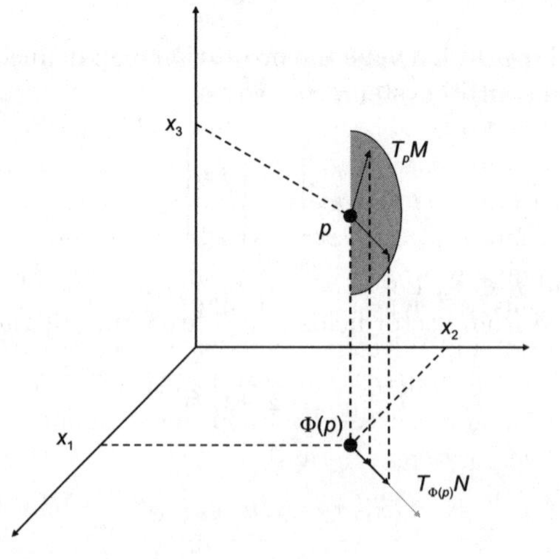

Fig. 5.5. The mapping of vector fields in S_M to S_N

Furthermore, reachability of a set can be defined for $A \subseteq M$ as

$$\text{Reach}(A, S) = \bigcup_{p \in A} \text{Reach}(p, S). \tag{5.9}$$

Using the concept of reachability, controllability can be restated as follows.

Definition 5.10. ***Controllability:*** *A control system $S = (B, F)$ is controllable if for every $p \in M$, $\text{Reach}(p, S) = M$.*

One of the important results from [47] is the following theorem that provides a condition for an abstracted system S_N to be controllable.

Theorem 5.11. ***Reachability for Φ-Related Systems:*** *Consider two control systems $S_M = (B_M, F_M)$ and $S_N = (B_N, F_N)$ which are Φ-related with respect to some surjective submersion $\Phi : M \to N$. Then for all $p \in M$,*

$$\Phi(\text{Reach}(p, S_M)) \subseteq \text{Reach}(\Phi(p), S_N). \tag{5.10}$$

Thus it follows that if S_M is controllable then S_N is controllable.

5.2 Consistent Control System Abstractions

Theorem 5.11 allows us to propagate controllability from the original system to abstracted system. But it is of more interest to study the propagation of controllability in the reverse direction. If the relationship is known, then the controllability of the original system can be determined from that of abstracted one. The definitions and theorems presented in this section are from [47] and are needed to study the propagation of controllability in the reverse direction. The complete proofs to these theorems are given in [47].

Definition 5.12. ***Implementability:*** *Let $\Phi : M \to N$ be a smooth surjection. A control system $S_M = (B_M, F_M)$ is implementable by control system $S_N = (B_N, F_N)$ if whenever there exists a trajectory in S_N connecting $q_1, q_2 \in N$, then there exist points $p_1 \in \Phi^{-1}(q_1)$ and $p_2 \in \Phi^{-1}(q_2)$ and a trajectory in S_M connecting them.*

Implementability ensures the existence of trajectories in the original system for every trajectory in the abstracted system. It is dependent upon particular elements chosen from the equivalence classes in original system corresponding to elements of abstracted one. Implementability does *not* guarantee the existence of trajectories between all elements of the equivalence classes.

Theorem 5.13. ***Implementability Condition:*** *Let $\Phi : M \to N$ be a smooth surjection. A control system $S_M = (B_M, F_M)$ is implementable by control system $S_N = (B_N, F_N)$ if and only if for all $q \in N$, $\text{Reach}(q, S_N) \subseteq \Phi(\text{Reach}(\Phi^{-1}(q), S_M))$. For implementability of Φ-related systems, the inclusion becomes equality.*

We have seen that implementability depends upon a particular element chosen from an equivalence class. In order to propagate controllability from the abstracted system to the original system, this dependence must be removed. To remove the dependence, the concept of consistency, as defined below, is used.

Definition 5.14. ***Consistency:*** *Let $\Phi : M \to N$ be a smooth surjection. A control system $S_M = (B_M, F_M)$ is consistent with respect to Φ whenever the following holds. If there exists a trajectory in S_M connecting $p_1, p_2 \in M$, then for all $p_1' \in \Phi^{-1}(\Phi(p_1))$ and $p_2' \in \Phi^{-1}(\Phi(p_2))$ there exists a trajectory in S_M connecting p_1' to p_2'.*

As with implementability, a condition for consistency can be stated in terms of reachability.

Theorem 5.15. ***Consistency Condition:*** *Let $\Phi : M \to N$ be a smooth surjection. A control system $S_M = (B_M, F_M)$ is consistent with respect to Φ if and only if for all $p \in M$,* $\mathrm{Reach}(p, S_M) = \Phi^{-1}(\Phi(\mathrm{Reach}(\Phi^{-1}(\Phi(p)), S_M)))$.

When implementability and consistency are combined, it provides a powerful result for controllability in reverse direction. This is given in the following theorem.

Theorem 5.16. ***Controllability Equivalence:*** *Let $\Phi : M \to N$ be a smooth surjection and suppose control systems $S_M = (B_M, F_M)$ and $S_N = (B_N, F_N)$ are Φ-related. Furthermore, suppose that S_N is implementable by S_M and S_M is consistent with respect to Φ. Then S_N is controllable if and only if S_M is controllable.*

5.3 Application to the Robotic Car and Unicycle

In this section we apply the notion of Φ-relatedness to the car-like robot with respect to some aggregation map Φ. Consider the robotic car to be the complex control system. We want to find Φ such that the abstracted system is a unicycle. The car kinematic model was given in Section 3.2.4 as

$$\begin{bmatrix} \dot{x} \\ \dot{y} \\ \dot{\theta} \\ \dot{\phi} \end{bmatrix} = \begin{bmatrix} \cos\theta \\ \sin\theta \\ \frac{\tan\phi}{l} \\ 0 \end{bmatrix} v_1 + \begin{bmatrix} 0 \\ 0 \\ 0 \\ 1 \end{bmatrix} v_2 \tag{5.11}$$

and the unicycle model was given in Section 3.2.3 by

$$\begin{bmatrix} \dot{x} \\ \dot{y} \\ \dot{\theta} \end{bmatrix} = \begin{bmatrix} \cos\theta \\ \sin\theta \\ 0 \end{bmatrix} v + \begin{bmatrix} 0 \\ 0 \\ 1 \end{bmatrix} \omega. \tag{5.12}$$

Proposition 5.17. *The car and unicycle are Φ-related control systems with Φ defined for each $(x, y, \theta, \phi) \in M$ by*

$$\Phi(x, y, \theta, \phi) = (x, y, \theta). \tag{5.13}$$

Proof: Consider any point $(x_0, y_0, \theta_0, \phi_0) \in M$. Then

$$\pi_M^{-1}(x_0, y_0, \theta_0, \phi_0) = ((x_0, y_0, \theta_0, \phi_0), (v_1, v_2)). \tag{5.14}$$

Thus,

$$T\Phi \circ F_M(\pi^{-1}(x_0, y_0, \theta_0, \phi_0)) = ((x_0, y_0, \theta_0), (\dot{x}_0, \dot{y}_0, \dot{\theta}_0)). \tag{5.15}$$

For any v_1 and v_2, letting $v = v_1$ and $\omega = \dot{\theta}_0$ gives

$$(x_0, y_0, \theta_0) = (x_u, y_u, \theta_u) \in F_N(\pi_N^{-1}(\Phi(x_0, y_0, \theta_0, \phi_0))). \tag{5.16}$$

So

$$T\Phi \circ F_M(\pi_M^{-1}(x_0, y_0, \theta_0, \phi_0)) \subseteq F_N(\pi_N^{-1}(\Phi(x_0, y_0, \theta_0, \phi_0))). \tag{5.17}$$

■

Proposition 5.18. *The car is consistent with respect to the Φ given by (5.13).*

Proof: Let $p_1 = (x_1, y_1, \theta_1, \phi_1)$ and $p_2 = (x_2, y_2, \theta_2, \phi_2)$ be two points in M with a trajectory in S_M connecting them. Let $\hat{v}_1(t)$ and $\hat{v}_2(t)$ be the control inputs defined for $0 \leq t \leq T$ that generate that trajectory. Then for any $\phi_0, \phi_3 \in (-\pi/2, \pi/2)$, the points $p_1' = (x_1, y_1, \theta_1, \phi_0)$ and $p_2' = (x_2, y_2, \theta_2, \phi_3)$ are in $\Phi^{-1}(\Phi(p_1))$ and $\Phi^{-1}(\Phi(p_2))$ respectively. For any p_1' and p_2', define the control inputs by:

$$v_1(t) = \begin{cases} 0 & t \in [0, 1) \\ \hat{v}_1(t-1) & t \in [1, T+1) \\ 0 & t \in [T+1, T+2] \end{cases} \tag{5.18}$$

$$v_2(t) = \begin{cases} \phi_1 - \phi_0 & t \in [0, 1) \\ \hat{v}_2(t-1) & t \in [1, T+1) \\ \phi_3 - \phi_2 & t \in [T+1, T+2] \end{cases} \tag{5.19}$$

These control inputs generate trajectories between p_1' and p_2'. ■

Proposition 5.19. *The unicycle is implementable by the car.*

Proof: Consider any two points $(x_1, y_1, \theta_1), (x_2, y_2, \theta_2) \in N$ with a trajectory between them. The points $(x_1, y_1, \theta_1, 0), (x_2, y_2, \theta_2, 0) \in M$ are in $\Phi^{-1}(x_1, y_1, \theta_1)$ and $\Phi^{-1}(x_2, y_2, \theta_2)$ respectively. Since the car is controllable, there is a trajectory in S_M connecting the two points. ■

5.4 Traceable Control Systems

The ideas of abstraction must be extended in order to **design a controller** in the simpler abstracted system and then transform the inputs back to the original system. In doing this, one must ensure that the transformed inputs cause the original system to behave as desired. The results from [47] reviewed in Section 5.2 give conditions (implementability and consistency) for the equivalence of controllability between a control system and its abstraction. Knowing this, however, is not enough to guarantee the existence of an input transformation that makes the original system track the abstracted system. The concepts of implementability and consistency, while important in determining the equivalence of the two systems, *do not solve the control design problem.* In particular, the following possibilities must be addressed:

1. Do there exist inputs in the abstracted system for which the transformation is not defined?
2. Do there exist inputs in the abstracted system whose transformation gives inputs in the original system defining a trajectory that does not correspond to the trajectory in the abstracted system?

In the case of the car/unicycle example, the conditions of implementability and consistency are met. However, there is an important aspect of the car/unicycle system that is not captured by these conditions. Given inputs for the unicycle, $v(t)$ and $\omega(t)$, $t \in [0, t_f]$, the input transformation given by

$$
\begin{aligned}
v_1(t) &= v(t) \\
v_2(t) &= \frac{l(v(t)\dot{\omega}(t) - \dot{v}(t)\omega(t))}{v^2(t) + l^2\omega^2(t)}
\end{aligned}
\tag{5.20}
$$

is well defined if $v(t) \neq 0$ and $\omega(t) \neq 0$ for all t, and both v and ω are differentiable. This transformation should provide the corresponding car trajectory if the initial conditions match, i.e., $x(0) = x_u(0)$, $y(0) = y_u(0)$, $\theta(0) = \theta_u(0)$, and $\phi(0) = \tan^{-1}(l\omega(0)/v(0))$. However, if $v = 0$ and ω is constant, then the resulting car inputs are $v_1 = 0$ and $v_2 = 0$, and the initial steering angle is $\phi(0) = \pm\pi/2$. In this case, the transformation is well defined but the car's initial condition lies at a point of singularity and the resulting trajectory fails to follow the unicycle trajectory (the car does not move while the unicycle is rotating). In fact, the derivative of θ becomes $\infty \cdot 0$ and thus is indeterminate. This shows that there are surfaces in the car's manifold, defined by $\dot{x} = \dot{y} = 0$ and $\dot{\theta} \neq 0$ to which vector fields cannot be tangential. The surface S in Figure 5.6 demonstrates this idea. Vector fields can point onto and out of such a surface (f_1 and f_2), but not along it (f_3), i.e., the car can travel to the surface and away from it, but not on it.

The car/unicycle example gives evidence that point #2 stated above is a real possibility in an actual system. This phenomena happens because the car's steering angle *cannot* be $\pi/2$. However, there do exist car inputs that result in the car's trajectory being arbitrarily close to the rotating unicycle. This condition is

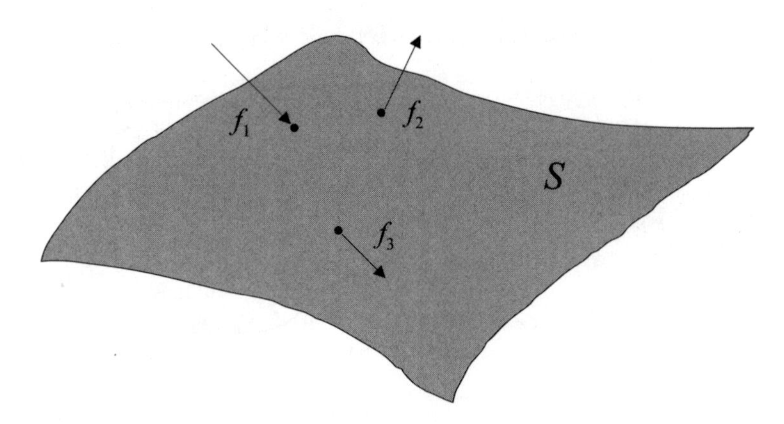

Fig. 5.6. A surface in the car's manifold

called ϵ-traceability. In the case of the car example, if the initial steering angle is chosen close to $\pi/2$ and held fixed while choosing $v_1(t) = l\dot{\theta}_u(t)/\tan\phi(0)$, then the angular velocity of the car, $\dot{\theta}$, is finite and matches the unicycle angular velocity exactly, as long as the initial conditions match. In terms of Figure 5.6, the unicycle can travel along the surface S while the car cannot. However, the car can reach S and remain arbitrarily close to it.

The importance of this example can be seen in the context of motion planning, where the controller is designed for the unicycle and then must be transformed by (5.20) into the car's inputs. Suppose the path shown in Figure 5.7 is given for the car to follow. This path is smooth everywhere except at point p_0. If a controller for the unicycle is designed to follow this path, the unicycle will stop and rotate when it arrives at p_0. However, as discussed above, the car cannot do this maneuver. The car's inputs can be determined everywhere, except at p_0, using the transformation given in (5.20). If the car inputs are determined offline, then an optimal trajectory about p_0 can be calculated.

As an example, consider the cost function given by

$$J = \int_{t_0}^{t_f} \left[w_1 \left([x(t) - x_u(t)]^2 + [y(t) - y_u(t)]^2 \right) + w_2 \left[\theta(t) - \theta_u(t) \right]^2 \right] dt \quad (5.21)$$

where w_1 and w_2 are weights determining the relative importance of each error term. For a smooth path, the transformation (5.20) minimizes J since x, y, and θ match x_u, y_u, and θ_u respectively. For the nonsmooth trajectory in Figure 5.7, no car inputs exist to minimize J, but J can be made arbitrarily small since the unicycle is ϵ-traceable by the car. Two possible car trajectories, Γ_1 and Γ_2, are shown in Figure 5.8 that depend on the weights in (5.21) and the maximum allowed value for J.

In the case of feedback control, where both the unicycle inputs and transformation to the car inputs are calculated online, a priori knowledge of the point p_0 is not available. The usefulness of ϵ-traceability is seen when the car reaches p_0 and the unicycle begins to rotate. We know from the above discussion that the

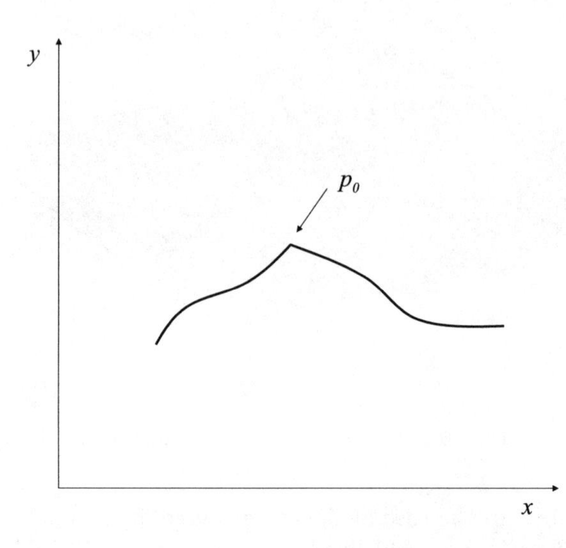

Fig. 5.7. A unicycle path that the car cannot follow

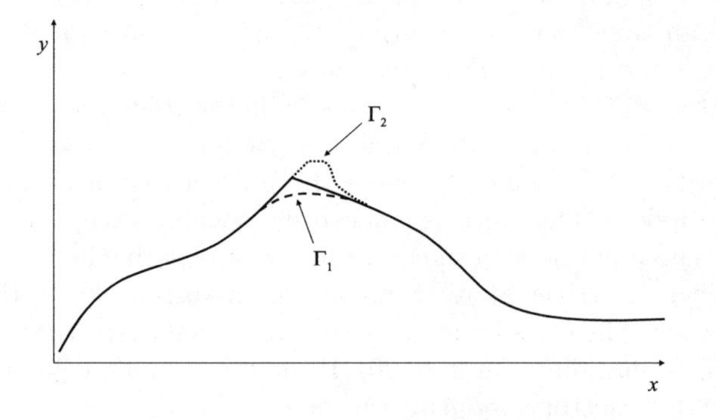

Fig. 5.8. Γ_1 and Γ_2 are possible paths that the car can follow

input transformation does not give the desired results. However, at p_0, the car can follow the rotating unicycle arbitrarily closely. One approach to determining the car inputs at p_0 is to allow the unicycle to finish its rotation and then calculate v_1 and v_2 so that the car describes a small circle starting at p_0 and ending with x, y, and θ matching x_u, y_u, and θ_u. Another approach is to drive the car so that its angle matches the unicycle's angle throughout the rotation, resulting in an offset between the car's (x, y) position and the unicycle's (x_u, y_u) position which then must be corrected. The choice of which approach to follow depends on the specifications of the particular problem. For example, is it more important that the car's angle match the unicycle's or can the angle error grow as long as the car's position remains close the to unicycle's? The answer to this question will determine the values for w_1 and w_2 in (5.21) and the resulting car inputs.

With this motivation, the following definitions are presented to precisely formulate the existence of these trajectories.

Definition 5.20. ***Traceability:*** *Consider Φ-related control systems $S_M = (B_M, F_M)$ and $S_N = (B_N, F_N)$ and a smooth surjection $\Phi : M \to N$. Then S_N is traceable by S_M if and only if for every trajectory c_N of S_N there exists a trajectory in S_M, $c_M : I \to M$, such that $\Phi(c_M(t)) = c_N(t)$ for all $t \in I$.*

Traceability is a stronger condition than implementability since traceability requires that there exists a trajectory in S_M for which the entire trajectory maps to the trajectory in S_N. Implementability only requires that there exists a trajectory in S_M for which its endpoints map to the endpoints of the trajectory in S_N. Any control system, S_N, that is traceable by S_M is implementable by S_M. However, the converse does not hold. The unicycle is an example of a system that is implementable by the car, but not traceable. However, as the above development shows, the unicycle is *almost* traceable by the car. This gives rise to the idea of ϵ-traceability. But first we define the distance between trajectories and give an example.

Definition 5.21. ***Distance Between Trajectories:*** *Let $S_N = (B_N, F_N)$ be a control system defined on Riemannian manifold $(N, \mathbf{g})$ with metric tensor $\mathbf{g} = (g_{ij})$ and let $c_1, c_2 : I \to N$ be trajectories in S_N. Then the distance between c_1 and c_2 is defined by the Poincare distance*

$$d_P(c_1, c_2) = \sup_{t\in I} \inf_{\tau \in I} d(c_1(t), c_2(\tau)) \tag{5.22}$$

The distance between points $d(\cdot,\cdot)$, which is invariant with respect to changes in coordinate systems, is given as

$$d(q_1, q_2) = \inf \int_a^b \left| g_{ij} \frac{dx_i}{dt} \frac{dx_j}{dt} \right|^{1/2} dt \tag{5.23}$$

where $(x_1, \ldots, x_n)$ are the local coordinates and the infimum is taken over all curves $\gamma(t)$ in N such that $\gamma(a) = q_1$ and $\gamma(b) = q_2$.

For the following development, car trajectories will be denoted by $X(t) = (x(t), y(t), \theta(t), \phi(t))$ and unicycle trajectories by $Y(t) = (x_u(t), y_u(t), \theta_u(t))$. The following proposition shows that for a unicycle trajectory, c_N, that has no corresponding car trajectory, c_M, with $c_N(t) = \Phi(c_M(t))$, there does exist a car trajectory with $\Phi(c_M(t))$ arbitrarily close to $c_N(t)$.

Proposition 5.22. *Given a unicycle trajectory $Y(t) = (\alpha, \beta, \theta_u(t))$, where α and β are constants (i.e., $\dot{x}_u = \dot{y}_u = 0$), and given any $\epsilon > 0$, there exists a car trajectory $X(t)$ such that $d_P(Y(t) - \Phi(X(t))) < \epsilon$.*

Proof: Consider any $\epsilon > 0$. Choose the car's initial conditions as $x(0) = \alpha$, $y(0) = \beta$, $\theta(0) = \theta_u(0)$, and $\phi(0) \in (-\pi/2, \pi/2)$ so that $\tan\phi(0) > 2l/\epsilon$. Let $v_1(t) = l\dot{\theta}_u(t)/\tan\phi(0)$ and $v_2(t) = 0$. The car's resulting trajectory $X(t)$ is

$$x(t) = \alpha + \frac{l}{\tan \phi(0)} \sin \theta_u(t) \tag{5.24}$$
$$y(t) = \beta + \frac{l}{\tan \phi(0)} (1 - \cos \theta_u(t)) \tag{5.25}$$
$$\theta(t) = \theta_u(t) \tag{5.26}$$
$$\phi(t) = \phi(0). \tag{5.27}$$

By choice of ϕ, $|x(t) - x_u(t)| < \epsilon$ and $|y(t) - y_u(t)| < \epsilon$ for all t. Hence $d_P(Y(t) - \Phi(X(t))) < \epsilon$. ■

The concept of trajectories in S_M that map arbitrarily close to those in S_N is formalized in the following definition of ϵ-traceability.

Definition 5.23. ***ϵ-Traceability:*** *Let $S_M = (B_M, F_M)$ and $S_N = (B_N, F_N)$ be Φ-related control systems and let $\Phi : M \to N$ be a smooth surjection. Then S_N is ϵ-traceable by S_M if given $\epsilon > 0$ and a trajectory $c_N : I \to N$ then there exists a trajectory $c_M : I \to M$ such that $d_P(c_N - \Phi(c_M)) < \epsilon$.*

An immediate consequence of this definition is that S_N is ϵ-traceable by S_M for $\epsilon = 0$ if and only if S_N is traceable by S_M. Let c_N be a trajectory in N. When $\epsilon = 0$, the definition for traceable is satisfied since there exists a trajectory c_M in M with $d_P(\Phi(c_M) - c_N) = 0$. By definition of the distance in N, it is necessary that $\Phi(c_M) = c_N$. Furthermore, if S_N is traceable by S_M, then there exists a trajectory c_M with $\Phi(c_M) = c_N$. Thus $d_P(\Phi(c_M) - c_N) = 0$.

With ϵ-traceability, we have a weaker condition on the relationship between the Φ-related systems S_M and S_N than implementability gives. While implementability requires that there exist a trajectory in S_M that maps the endpoints of a given trajectory in S_N, ϵ-traceability requires that there exist a trajectory in S_M that can map arbitrarily close to the given trajectory in S_N. If S_N is ϵ-traceable by S_M, a condition stronger than consistency is needed for controllability to propagate between the two systems. A restriction is placed on the system to require that if there are converging sequences of trajectories in S_N then a corresponding trajectory exists in S_M. This leads to the following definition of ϵ-consistency stated precisely below.

Definition 5.24. ***ϵ-Consistency:*** *Let $S_M = (B_M, F_M)$ be a control system on M and let $\Phi : M \to N$ be a smooth surjection. Then S_M is ϵ-consistent with respect to Φ whenever the following holds. If for any $p_1, p_2 \in M$ there exist sequences p_{1_n} and p_{2_n} in M such that*

1. *there exist trajectories connecting $\Phi(p_{1_n})$ and $\Phi(p_{2_n})$ for all n and*
2. *$\Phi(p_{1_n}) \to \Phi(p_1)$ and $\Phi(p_{2_n}) \to \Phi(p_2)$,*

then there is a trajectory connecting p_1 and p_2.

Unlike consistency, which requires that trajectories exist between all members of $\Phi^{-1}(\Phi(p_1))$ and $\Phi^{-1}(\Phi(p_2))$ based on the existence of only one trajectory, ϵ-consistency gives the condition for a trajectory to exist between p_1 and p_2. Consistency does not involve trajectories in S_N, but for ϵ-consistency, the existence of a single trajectory relies upon a sequence of trajectories in S_N whose endpoints map arbitrarily close to $\Phi(p_1)$ and $\Phi(p_2)$. The following example shows that S_M can be ϵ-consistent with respect to Φ even if it is not controllable.

Example 5.25. Consider the system in S_M given by

$$\begin{bmatrix} \dot{x}_1 \\ \dot{x}_2 \end{bmatrix} = \begin{bmatrix} u \\ 0 \end{bmatrix} \tag{5.28}$$

and the system in S_N given by

$$\dot{x}_2 = 0 \tag{5.29}$$

with the Φ-map given by $\Phi(x_1, x_2) = x_2$. For $p_1 = (x_1, \alpha)$ and $p_2 = (x_1', \beta)$, let $p_{1_n} = \alpha$ and $p_{2_n} = \beta$ for all n. If $\alpha = \beta$, then there exist trajectories in S_N connecting $\Phi(p_{1_n})$ and $\Phi(p_{2_n})$ for all n and a trajectory in S_M connecting p_1 and p_2. However, if $\alpha \neq \beta$, there is no trajectory in S_M connecting p_1 and p_2 and no trajectories in S_N connecting $\Phi(p_{1_n})$ and $\Phi(p_{2_n})$.

Controllability of S_M is thus not a necessary condition for ϵ-consistency. However, the following corollary shows that Φ-relatedness between S_M and S_N and controllability of S_M are sufficient conditions for S_M to be ϵ-consistent with respect to Φ.

Corollary 5.26. *Consider control systems $S_M = (B_M, F_M)$ and $S_N = (B_N, F_N)$ which are Φ-related with respect to smooth surjection $\Phi : M \to N$. If S_M is controllable, then S_M is ϵ-consistent with respect to Φ.*

Proof: Consider any $p_1, p_2 \in M$. Let $p_{1_n} = p_1$ and $p_{2_n} = p_2$ for all n. Since S_M is controllable, there exist trajectories connecting p_{1_n} and p_{2_n} for all n. Denote each such trajectory by c_M. Then $\Phi(c_M)$ is also a trajectory in S_N. The sequences p_{1_n} and p_{2_n} satisfy the conditions of Definition 5.24. ■

Thus it follows that the car is ϵ-consistent with respect to the Φ mapping defined by (5.13).

As with implementability and consistency, the relationship between ϵ-traceability and ϵ-consistency provide a means to propagate controllability between control systems S_M and S_N.

Theorem 5.27. *Controllability Equivalence:* *Consider control systems $S_M = (B_M, F_M)$ and $S_N = (B_N, F_N)$ which are Φ-related with respect to smooth surjection $\Phi : M \to N$. Assume that S_N is ϵ-traceable by S_M and S_M is ϵ-consistent with respect to Φ. Then S_N is controllable if and only if S_M is controllable.*

Proof: If S_M is controllable then S_N is controllable since the control systems are Φ-related. Now assume S_N is controllable and consider any points $p_1, p_2 \in M$. Because S_N is controllable, there is a trajectory c_N connecting $\Phi(p_1)$ and $\Phi(p_2)$. Then for each n, ϵ-traceability gives a sequence of trajectories c_{M_n} such that $d_P(c_N - \Phi(c_{M_n})) < 1/n$. Denote the endpoints of each c_{M_n} by p_{1_n} and p_{2_n}. These endpoints provide sequences in M with $\Phi(p_{1_n}) \to \Phi(p_1)$ and $\Phi(p_{2_n}) \to \Phi(p_2)$. Because S_N is controllable, there exist trajectories connecting $\Phi(p_{1_n})$ and $\Phi(p_{2_n})$ for all n. Since S_M is ϵ-consistent, there is a trajectory connecting p_1 and p_2. Thus S_M is controllable. ■

With these notions defined, there is now a framework in which to approach the control design for the car. In the following chapters, control design using abstraction is investigated in detail and applied to the car/robot system. The current chapter concludes with a comparison between traceability and differential flatness, an equivalence relationship between systems.

5.5 Traceability and Differential Flatness

The notion of differential flatness was developed in a series of papers by Fliess et al [14][15][16]. Differential flatness can be summarized by the following characteristics:

- Differential flatness deals with equivalence of systems.
- The meaning of equivalence is that there is a local one-to-one correspondence between trajectories of different systems and that the variables of one system can be expressed as a function of the variables of the other and a finite number of their derivatives.
- Differential flatness borrows the notion of Lie-Bäcklund isomorphism from physics.
- The size of the state space is not preserved and system complexity may be reduced.
- The number of inputs is invariant.
- A system is differentially flat if it is Lie-Bäcklund equivalent to a trivial system (a system without dynamics).

The following is a summary of ideas put forth in [16]. The notion of differential flatness was motivated by dynamic feedback linearization. Given a nonlinear system

$$\dot{x} = f(x, u) \tag{5.30}$$

where $x \in \mathbb{R}^n$ and $u \in \mathbb{R}^m$, the goal of dynamic feedback linearization is to find a dynamic compensator of the form

$$\dot{\omega} = a(x, \omega, v) \tag{5.31}$$

$$u = \alpha(x, \omega, v) \tag{5.32}$$

$$z = \phi(x, \omega) \tag{5.33}$$

with $\omega \in \mathbb{R}^q$ and $v \in \mathbb{R}^m$ such that

$$\dot{z} = Az + Bv \tag{5.34}$$

where $z \in \mathbb{R}^{n+q}$. Although flatness is closely related to dynamic feedback linearization, it is a distinct system property with other applications (e.g., motion planning, tracking). The system (5.30) is said to be *differentially flat* if there exists a finite set of differentially independent variables $y = (y_1, \ldots, y_m)$ (called the linearizing or flat outputs) such that

1. These y_i's are functions of the system variables and a finite number of their derivatives.
2. Any system variable is a function of the y_i's and a finite number of their derivatives.

In other words, if (5.30) is differentially flat, then there exist m flat outputs with

$$y = h(x, u_1, \ldots, u_1^{(\beta_1)}, \ldots, u_m, \ldots, u_m^{(\beta_m)}) \tag{5.35}$$

$$x = g_1(y_1, \ldots, y_1^{(\alpha_1)}, \ldots, y_m, \ldots, y_m^{(\alpha_m)}) \tag{5.36}$$

$$u = g_2(y_1, \ldots, y_1^{(\alpha_1+1)}, \ldots, y_m, \ldots, y_m^{(\alpha_m+1)}). \tag{5.37}$$

From a differential geometric point of view, the vector field f in (5.30) can be seen as a countably infinite set of vector fields parameterized by u. A trajectory in the system, typically determined by $x(0)$ and $u(t)$ for $t \in [0, T]$, can instead be represented by a single infinite dimensional point $\xi_0 = (x(0), u(0), \dot{u}(0), \ddot{u}(0), \ldots, u^{(\mu)}(0), \ldots)$. A "prolongation" of f is given by

$$F(\xi) = (f(x, u), \dot{u}, \ddot{u}, \ldots) \tag{5.38}$$

where $\xi \in X \times U \times \mathbb{R}_m^\infty$ and $\mathbb{R}_m^\infty = \mathbb{R}_m \times \mathbb{R}_m \times \cdots$. Then the system (5.30) can be represented by

$$\dot{\xi} = F(\xi). \tag{5.39}$$

Using this notation, a *classic system* is defined in [16] by the pair $(X \times U \times \mathbb{R}_m^\infty, F)$, with F being a smooth vector field on $X \times U \times \mathbb{R}_m^\infty$. A *trivial system* is defined by $(\mathbb{R}_m^\infty, F_m)$ where the coordinates are $y = (y_1, \ldots, y_m)$ and its derivatives $y^{(1)} = (y_1^{(1)}, \ldots, y_m^{(1)})$, $y^{(2)} = (y_1^{(2)}, \ldots, y_m^{(2)})$. The vector field is

$$F_m(y, y^{(1)}, y^{(2)}, \ldots) = (y, y^{(1)}, y^{(2)}, \ldots), \tag{5.40}$$

a chain of m independent integrators.

An equivalence relation between systems is informally described as an invertible transformation exchanging trajectories of systems. For Φ-related vector fields F and G, the relation

$$G(\Phi(\xi)) = \frac{\partial \Phi}{\partial \xi}(\xi) F(\xi) \tag{5.41}$$

implies that $\Phi(\xi(t))$ is a trajectory. The transformation Φ is called an *endogenous transformation* if it has a smooth inverse.

Formally, two systems are *differentially equivalent* at a pair of points (p, q) if there exists a smooth mapping Φ from a neighborhood of p to a neighborhood of q that is an endogenous transformation. Using this definition, differential flatness can be defined as follows: A control system is *differentially flat* about p if it is equivalent to a trivial system in a neighborhood of p. If this definition is satisfied for all points in the state space, then the system is differentially flat. This idea can be extended to the time-varying case if endogenous transformations are replaced with Lie-Bäcklund isomorphisms. In such a case, the system is called *orbitally flat*.

In summary, differentially flat systems are equivalent to systems without dynamics (trivial systems) through an endogenous transformation. The flat outputs contain all of the dynamic information about a system and the system states and inputs can be found algebraically as a function of flat outputs and their derivatives.

Differential flatness has application to motion planning and tracking. This can easily be seen in the example of the car with n trailers. The system is shown in Figure 5.9[49] and the dynamics are given by

$$\begin{aligned}
\dot{x}_0 &= u_1 \cos\theta_0 \\
\dot{y}_0 &= u_1 \sin\theta_0 \\
\dot{\phi} &= u_2 \\
\dot{\theta}_0 &= \frac{u_1}{d_0} \tan\phi \\
\dot{\theta}_i &= \frac{u_1}{d_0} \left(\prod_{j=1}^{i-1} \cos(\theta_{j-1} - \theta_j) \right) \sin(\theta_{j-1} - \theta_j) \text{ for } i = 1, \ldots, n
\end{aligned} \tag{5.42}$$

where $x_0, y_0 \in \mathbb{R}$ is the position of the car, and $u_1, u_2 \in \mathbb{R}$ are the inputs. The flat outputs are x_n and y_n, the position of the last trailer given by

$$x_n = x_0 - \sum_{i=1}^{n} d_i \cos\theta_i \tag{5.43}$$

$$y_n = y_0 - \sum_{i=1}^{n} d_i \sin\theta_i . \tag{5.44}$$

Thus, the flat outputs are a function of the states only. It has been shown [49] that given the trajectory of $x_n(t)$ and $y_n(t)$, the inputs $u_1(t)$ and u_2 can be found explicitly.

Differential flatness provides a characterization of system equivalence that is different from ϵ-traceability in two ways. First, differential flatness relies upon a local one-to-one correspondence between trajectories of system. Second, differential flatness requires that the number of inputs be the same for each system.

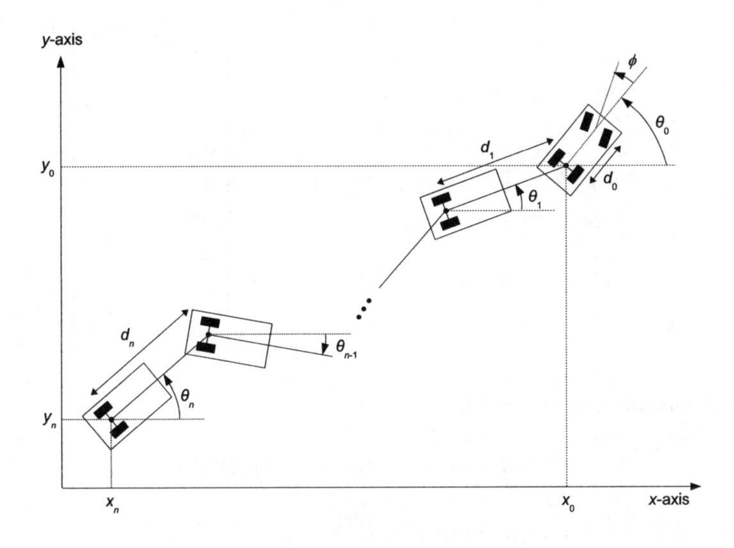

Fig. 5.9. The car with n trailers. Adapted from [49].

The concept of ϵ-traceability does not have these requirements. Here an example is provided of an underwater vehicle abstracted to a rolling disk. In this case, the rolling disk is not traceable by the underwater vehicle, but it is ϵ-traceable, and the number of inputs is not invariant.

Consider the nonholonomic model of an underwater vehicle [40]

$$\begin{bmatrix} \dot{x} \\ \dot{y} \\ \dot{z} \\ \dot{\phi} \\ \dot{\theta} \\ \dot{\psi} \end{bmatrix} = \begin{bmatrix} \cos\theta\cos\psi \\ \cos\theta\sin\psi \\ -\sin\theta \\ 0 \\ 0 \\ 0 \end{bmatrix} v + \begin{bmatrix} 0 \\ 0 \\ 0 \\ 1 \\ 0 \\ 0 \end{bmatrix} \omega_x + \begin{bmatrix} 0 \\ 0 \\ 0 \\ \sin\phi\tan\theta \\ \cos\phi \\ \sin\phi\sec\theta \end{bmatrix} \omega_y + \begin{bmatrix} 0 \\ 0 \\ 0 \\ \cos\phi\tan\theta \\ -\sin\phi \\ \cos\phi\sec\theta \end{bmatrix} \omega_z \quad (5.45)$$

where x, y, and z are the Cartesian coordinates and ϕ, θ, and ψ are the Euler angles about the (vehicle) x, y, and z-axes, respectively. The inputs to the system are v, the linear velocity along the vehicle's x-axis, and ω_x, ω_y, and ω_z, the rotational velocities about the vehicle axes. Note that there is a singularity at $\theta = \pi$.

This model can be modified to represent an underwater vehicle whose angular velocities cannot be controlled directly. Suppose the vehicle's yaw and pitch are controlled by rudders with angles β_1 and β_2 located at $-l_1$ and $-l_2$ in the vehicle frame. Then $\omega_x = 0$, $\omega_y = v(\tan\beta_2)/l_2$, and $\omega_z = -v(\tan\beta_1)/l_1$. The resulting model is

$$
\begin{bmatrix} \dot{x} \\ \dot{y} \\ \dot{z} \\ \dot{\phi} \\ \dot{\theta} \\ \dot{\psi} \\ \dot{\beta}_1 \\ \dot{\beta}_2 \end{bmatrix} = \begin{bmatrix} \cos\theta\cos\psi \\ \cos\theta\sin\psi \\ -\sin\theta \\ \tan\theta\left(\sin\phi\frac{\tan\beta_2}{l_2} - \cos\phi\frac{\tan\beta_1}{l_1}\right) \\ \cos\phi\frac{\tan\beta_2}{l_2} + \sin\phi\frac{\tan\beta_1}{l_1} \\ \sec\theta\left(\sin\phi\frac{\tan\beta_2}{l_2} - \cos\phi\frac{\tan\beta_1}{l_1}\right) \\ 0 \\ 0 \end{bmatrix} v + \begin{bmatrix} 0 \\ 0 \\ 0 \\ 0 \\ 0 \\ 0 \\ 1 \\ 0 \end{bmatrix} \omega_1 + \begin{bmatrix} 0 \\ 0 \\ 0 \\ 0 \\ 0 \\ 0 \\ 0 \\ 1 \end{bmatrix} \omega_2. \tag{5.46}
$$

If the Φ-mapping is defined by

$$
\Phi(x, y, z, \phi, \theta, \psi, \beta_1, \beta_2) = (x, y, \psi), \tag{5.47}
$$

then this gives the projection of the underwater vehicle onto the x, y plane, whose dynamics are

$$
\begin{bmatrix} \dot{x} \\ \dot{y} \\ \dot{\psi} \end{bmatrix} = \begin{bmatrix} \cos\psi \\ \sin\psi \\ 0 \end{bmatrix} v_u + \begin{bmatrix} 0 \\ 0 \\ 1 \end{bmatrix} \omega_u \tag{5.48}
$$

where $v_u = v\cos\theta$ and $\omega_u = \sec\theta\,[(\sin\phi)(\tan\beta_2)/l_2 - (\cos\phi)(\tan\beta_1)/l_1]$. If v_u and ω_u are independent inputs, then these dynamics match the rolling disk.

The pair of systems (5.46) and (5.48) exhibit a relationship similar to the car and rolling disk. The rolling disk is not traceable by (5.46) because of the rolling disk's ability to rotate. The rolling disk is ϵ-traceable by (5.46) because the latter can turn in an arbitrarily small circle. However, the number of inputs in each system does not match. In this case, the framework of differential flatness cannot be used.

6 Control Design

This chapter focuses on control design using abstraction. Given a system and its abstraction, as defined in the previous chapter, a method is presented for transforming a controller in the abstracted system back to the original system. Conditions for the existence and uniqueness of transformation are provided and the relationship with traceability is given. Finally, if such a transformation does not exist, the existence of an arbitrarily close solution is investigated and its relationship to ϵ-traceability is studied.

6.1 Problem Statement

Let $\Phi : M \to N$ be a smooth mapping and let $S_M = (B_M, F_M)$ and $S_N = (B_N, F_N)$ be two Φ-related control systems. It will be assumed that for some fixed time t, the state of S_M is given as $p \in M$ and the control input for S_N is known. The point $\Phi(p) \in N$ and the control input in S_N, $v \in U_N$ uniquely define a vector field $Y(\Phi(p), v) \in T_{\Phi(p)}N$.

The goal for the control design is as follows. Find an input for S_M, $u = u(p, v) \in U_M$ for which the corresponding tangent vector $X(p, u) \in T_pM$ maps to $T_{\Phi(p)}N$ such that $T\Phi(X(p, u)) = Y(\Phi(p), v)$. In other words, choose an input for the original system so that the resulting tangent vector matches the tangent vector in the abstracted system when mapped through $T\Phi$. This guarantees that the movement on N can be followed by the original system on M.

6.2 Existence and Uniqueness of the Input Transformation

For any $p \in M$, its local representation is given as $x = (x_1, \ldots, x_n) \in \mathbb{R}^n$. Furthermore, $\Phi(p)$ is represented as a m-dimensional vector $y = (y_1, \cdots, y_m) \in \mathbb{R}^m$. The dynamics of S_M and S_N are locally given by

$$\dot{x} = F(x, u) \tag{6.1}$$

$$\dot{y} = G(y, v) \tag{6.2}$$

P. Mellodge & P. Kachroo: Model Abstraction in Dynamical Systems, LNCIS 379, pp. 81–86, 2008.
springerlink.com

where $u = (u_1, \ldots, u_l)$ and $v = (v_1, \ldots, v_k)$. It is necessary to find u so that

$$T\Phi(F(x,u)) = G(y,v) \tag{6.3}$$

or in expanded form

$$\begin{aligned} T\Phi_1(F(x,u)) &= G_1(y,v) \\ T\Phi_2(F(x,u)) &= G_2(y,v) \\ &\vdots \\ T\Phi_m(F(x,u)) &= G_m(y,v). \end{aligned} \tag{6.4}$$

Since x is fixed, the left side of (6.4) is only a function of u and the right side of (6.4) is a vector of m numbers $b = (b_1, \ldots, b_m)^T = (G_1(y,v), \ldots, G_m(y,v))^T$. So there is a mapping (different for each x) $h : U_M \to T_{\Phi(x)}N$ with $h(u) = (h_1(u), \ldots, h_m(u))^T$ such that

$$\begin{aligned} h_1(u) &= b_1 \\ h_2(u) &= b_2 \\ &\vdots \\ h_m(u) &= b_m. \end{aligned} \tag{6.5}$$

Three possibilities exist for finding a solution to these m algebraic equations:

1. u exists and is unique
2. u exists and is not unique
3. u does not exist

For the general case, a solution exists if h is onto. A unique solution exists if h is one-to-one and onto.

6.2.1 Special Case

The solution to a set of nonlinear algebraic equations can be difficult to find and a general solution method does not exist. Also, the condition that h be onto is excessive since there is only a single tangent vector that must be mapped to for each $p \in M$. If the class of control systems in limited to those that are linear in u (or affine), the system (6.1) is of the form

$$\dot{x} = F_a(x) + \sum_{i=1}^{l} F_{b_i}(x) u_i. \tag{6.6}$$

Then since $T\Phi(f(x,u)) = \frac{d\Phi}{dx} f(x,u)$ is a linear map,

$$T\Phi\left(F_a(x) + \sum_{i=1}^{l} F_b(x) u_i\right) = T\Phi(F_a(x)) + \sum_{i=1}^{l} T\Phi(F_{b_i}(x)) u_i \tag{6.7}$$

and

$$\sum_{i=1}^{l} T\Phi(F_{b_i}(x))u_i = G(y,v) - T\Phi(F_a(x)). \tag{6.8}$$

In this case, (6.5) is of the form

$$h(u) = \hat{b} \tag{6.9}$$

where $\hat{b} = [G_1(y,v) - T\Phi_1(F_a(x)), \ldots, G_m(y,v) - T\Phi_m(F_a(x))]^T$. This is a set of linear algebraic equations and the mapping h becomes a linear transformation. Solutions to linear equations are a well-studied phenomenon in linear algebra and they are known to exist if h has certain properties. The following two theorems are key results from [17] identify conditions for which the necessary input u exists and for which it is unique.

Theorem 6.1. ***Existence of Solutions [17]:*** *Let V and W be vector spaces and let $h : V \to W$ be a linear transformation. If* $\text{rank}(h) = \dim(W)$*, then h is onto.*

This theorem applies to the above problem if the input space U_M is a vector space, i.e., for any valid system inputs u_1 and u_2, $\alpha u_1 + \beta u_2$ is also a valid system input for any real numbers α and β. Also, from Definition 2.10, it is known that $T_{\Phi(p)}N$ is a vector space. This theorem states that if the dimension of range space of h (rank of h) is m, then h is onto and a u exists to satisfy (6.9).

Theorem 6.2. ***Uniqueness of Solutions [17]:*** *Let V and W be vector spaces and let $h : V \to W$ be a linear transformation. Then h is one-to-one if and only if $N(h) = \{0\}$.*

In this theorem, $N(h)$ denotes the null space of h, the set $\{u \in U_M : h(u) = 0\}$. The solution, if it exists, is unique if $u = 0$ is the only point that h maps to the 0 tangent vector in $T_{\Phi(p)}N$.

These two theorems can be combined to provide a sufficient condition for existence and uniqueness of the solution.

Theorem 6.3. *Let V and W be vector spaces and let $h : V \to W$ be a linear transformation. Then the inverse transformation $h^{-1} : W \to V$ exists if* $\text{rank}(h) = \dim(W)$ *and $N(h) = \{0\}$.*

Since h is a linear transformation, it can be represented by an $m \times l$ matrix A. Then (6.9) becomes

$$\begin{bmatrix} a_{11} & \cdots & a_{1l} \\ \vdots & \ddots & \vdots \\ a_{m1} & \cdots & a_{ml} \end{bmatrix} \begin{bmatrix} u_1 \\ \vdots \\ u_l \end{bmatrix} = \begin{bmatrix} b_1 \\ \vdots \\ b_m \end{bmatrix} \tag{6.10}$$

or in more compact form

$$Au = b. \tag{6.11}$$

For affine systems, the problem is reduced to identifying certain properties of A. Thus, the above theorems can be transformed into statements about A.

Theorem 6.4. *Let* $Au = b$ *be a system of linear equations. Then the system has at least one solution if and only if* $\text{rank}(A) = \text{rank}(A|b)$*. Furthermore, the solution is unique if and only if* $\text{rank}(A) = l$*.*

Here rank() indicates the number of independent rows and columns. It is sufficient, but not necessary, for the transformation h to be onto, so Theorem 6.1 provides extra restrictions on the system. Theorem 6.4 gives exact conditions (necessary and sufficient) for the existence of the solution by incorporating into b the specific tangent vector to which h must map. If the augmented matrix $(A|b)$ only adds a dependent column to A, then a solution exists and if, in addition, the $m \times l$ matrix has rank l, then the solution is unique. This rank condition implies that if $m < l$, the solution cannot be unique. In such a case, more independent rows must be added to (6.10) if a unique solution is desired. These additional rows place more restrictions on u and may be derived from optimal control criteria for the system.

6.3 Application to the Car/Unicycle System

From the previous chapter, it is known that there exists a unicycle trajectory (rotation) that no car trajectory can map to through Φ. In this section, this relationship is investigated in terms of the above results. The mapping from the car's input space to the unicycle's tangent space is studied in this context.

Given the car's state, (x, y, θ, ϕ), the unicycle's state is (x, y, θ). Also, the uncycle's inputs, v and ω are known. The relationship between the tangent vectors that must be satisfied is

$$\begin{bmatrix} \cos\theta & 0 \\ \sin\theta & 0 \\ \frac{\tan\phi}{l} & 0 \end{bmatrix} \begin{bmatrix} u_1 \\ u_2 \end{bmatrix} = \begin{bmatrix} v\cos\theta \\ v\sin\theta \\ \omega \end{bmatrix}. \tag{6.12}$$

The matrix representation of the transformation h is then

$$A = \begin{bmatrix} \cos\theta & 0 \\ \sin\theta & 0 \\ \frac{\tan\phi}{l} & 0 \end{bmatrix}. \tag{6.13}$$

The rank of A is one since the right column is all zeros. Augmenting this matrix with the unicycle tangent vector gives

$$(A|b) = \begin{bmatrix} \cos\theta & 0 & v\cos\theta \\ \sin\theta & 0 & v\sin\theta \\ \frac{\tan\phi}{l} & 0 & \omega \end{bmatrix}. \tag{6.14}$$

To satisfy Theorem 6.4, the rightmost column must be all zeros or a multiple of the first column. Thus, a solution exists in two cases: (1) v and ω are both zero or (2) $(\tan\phi)v/l = \omega$. In either case, the solution is not unique since the matrix is not full rank.

For the rotating unicycle trajectory, $v = 0$ and $\omega = k \neq 0$, the resulting augmented matrix

$$(A|b) = \begin{bmatrix} \cos\theta & 0 & 0 \\ \sin\theta & 0 & 0 \\ \frac{\tan\phi}{l} & 0 & k \end{bmatrix} \tag{6.15}$$

has rank 2. Thus, no solution exists. However, if given the freedom to choose the point in the car's state space in the equivalence class that maps to the unicycle's state, the car's tangent vector can be chosen close to that for the unicycle. Choose x, y, and θ to match the unicycle's and for any $\epsilon > 0$, choose ϕ so that $\tan\phi = 2l/\epsilon$ and let $u_1 = l\omega/\tan\phi$ and $u_2 = 0$. Then

$$\begin{bmatrix} \cos\theta & 0 \\ \sin\theta & 0 \\ \frac{\tan\phi}{l} & 0 \end{bmatrix} \begin{bmatrix} u_1 \\ u_2 \end{bmatrix} = \begin{bmatrix} \cos\theta & 0 \\ \sin\theta & 0 \\ \frac{2}{\epsilon} & 0 \end{bmatrix} \begin{bmatrix} \frac{\epsilon\omega}{2} \\ 0 \end{bmatrix} = \begin{bmatrix} \frac{\epsilon\omega}{2}\cos\theta \\ \frac{\epsilon\omega}{2}\sin\theta \\ \omega \end{bmatrix} \tag{6.16}$$

whose vector components are within ϵ of $[0\ 0\ \omega]^T$. Since ϵ was arbitrary, the car's tangent vector can be made arbitrarily close to the unicycle's.

6.4 Connection with Traceability

Traceability and the existence of the bijection h for each point in the original system are closely related, but are not equivalent characterizations of the relationship between control systems. Traceability guarantees the existence of at least one trajectory in the original system that maps to each trajectory in the abstracted system. On the other hand, the existence of the bijection h for each point in the original system guarantees that it can match any instantaneous movement of the abstracted system.

Traceability does not imply the existence of an input transformation for every point in M. Suppose the some $p \in M$ is given and consider any trajectory $c_N(t)$ of S_N passing through $\Phi(p)$. Traceability would guarantee that there is a trajectory $c_M(t)$ with $\Phi(c_M(t)) = c_N(t)$, but c_M does not necessarily pass through p. So there may not be a u that forces the tangent vector at p to map to the tangent vector at $\Phi(p)$.

There are certain circumstances for which the converse is true, but extra conditions on Φ are required. This circumstance is formulated in the following theorem.

Theorem 6.5. ***Sufficient Condition for Traceability:*** *Let $\Phi : M \to N$ be a smooth surjection and let S_M and S_N be Φ-related control systems. Furthermore, let Φ be locally invertible with smooth inverse. If for every $p \in M$ the mapping $h : U_M \to T_{\Phi(p)}N$ defined by (6.5) has a unique solution, then S_N is traceable by S_M.*

Proof: Let $c_N(t)$ be any trajectory of S_N for $t \in I$ and consider any t. There exists a unique u and a point x such that

$$\Phi(x) = c_N(t) \tag{6.17}$$

and

$$T\Phi(F(x,u)) = c_N'(t) \tag{6.18}$$

where F denotes the local dynamics of S_M. There also exist neighborhoods of x and $c_N(t)$, denoted by W_M and W_N, respectively, in which Φ is invertible and the local dynamics of S_M and S_N are F and G, respectively. Let $\Psi : W_N \to W_M$ be the smooth inverse mapping of Φ. Define $T\Psi$ to be the map that associates $F(\hat{x}, \hat{u})$ with $c_N'(\hat{t})$ for $c_N(\hat{t}) \in W_N$. Then, by Theorem 5.7, $\Psi(c_N(\hat{t}))$ is a trajectory in S_M. ■

The idea behind abstraction is to model a system using a simpler model. The simpler model is created using the smooth map Φ and its push forward map $T\Phi$. Because the model becomes simpler, and in general lower dimensional, some information is lost in this mapping. To obtain complete information about the original system from the abstracted system, there must be additional restrictions on the Φ mapping. These restrictions are, in part seen in the above theorem, that the Φ mapping must be invertible in some way.

If the system is not traceable or if the input transformation does not exist, control design can be done in such a way that the original system approximates trajectories in the abstracted system. In the next chapter, open-loop control design is investigated using the car/unicycle system with a unicycle trajectory that the car cannot achieve.

7 Open-Loop Control Design

As discussed in the Chapter 5, the car and uncycle have very close relationship that allows controllability to propagate between the two systems. However, it was shown that the uniycle is not traceable by the car. This means that there exists a unicycle trajectory that does not the Φ-mapping of any car trajectory. The particular trajectory that causes problems is that of the rotating unicycle and addressing this problem in the open-loop setting is the focus of this chapter.

In this chapter, an open-loop optimal control algorithm is presented that utilizes the ϵ-traceability of the unicycle by the car. The algorithm is developed and simulation results are given for different initial car inputs. Their results are compared.

7.1 Input Transformation

Based on kinematic models for the car (3.10) and unicycle (3.1), an input transformation can be found to convert $v(t)$ and $\omega(t)$ into $v_1(t)$ and $v_2(t)$. By setting $\dot{\theta}_u(t)$ and $\dot{\theta}(t)$ equal, we obtain

$$\omega(t) = \frac{\tan\phi(t)}{l} v_1(t). \tag{7.1}$$

Letting $v_1(t) = v(t)$ and assuming $v(t) \neq 0$ for all t, this equation can be solved for ϕ to give

$$\phi(t) = \tan^{-1}\frac{l\omega(t)}{v(t)}. \tag{7.2}$$

To find v_2, differentiate ϕ with respect to time to get the transformation between the unicycle inputs and the car inputs

$$\begin{aligned} v_1(t) &= v(t) \\ v_2(t) &= \frac{l\left[v(t)\dot{\omega}(t) - \dot{v}(t)\omega(t)\right]}{v^2(t) + l^2\omega^2(t)}. \end{aligned} \tag{7.3}$$

If the initial conditions of the car match those of the unicycle, $x(0) = x_u(0)$, $y(0) = y_u(0)$, $\theta(0) = \theta_u(0)$, and an additional condition is met for the initial

P. Mellodge & P. Kachroo: Model Abstraction in Dynamical Systems, LNCIS 379, pp. 87–96, 2008.
springerlink.com

steering angle, $\phi(0) = \tan^{-1}[l\omega(0)/v(0)]$, then the input transformation given in (7.3) will cause the car's trajectory to match the unicycle's. In other words,

$$\begin{aligned} x(t) &= x_u(t) \\ y(t) &= y_u(t) \\ \theta(t) &= \theta_u(t) \end{aligned}$$

for all t for which the unicycle trajectory is defined.

It is immediately obvious that the tranformation (7.3) will fail if either v or ω is not differentiable or if both $v(t)$ and $\omega(t)$ are zero for any t. In these cases, the transformation is not well defined. However, there is an instance in which the transformation is well defined, but the resulting car inputs do not cause the car to track the unicycle. When $v = 0$ and $\omega \neq 0$, the unicycle is rotating about its axis. It is not possible for the car to perform this maneuver and the transformation results in both car inputs being zero. It is this latter situation that is the focus of this chapter.

7.2 Open-Loop Optimal Control

For the following development, it is assumed that the unicycle inputs, $v_1(t)$ and $v_2(t)$, are given for $t \in [0, t_f]$ and the resulting trajectory is specified by $x_u(t)$, $y_u(t)$, and $\theta_u(t)$. The cost function is given by

$$J = \frac{1}{2}\int_0^{t_f} \left[w_1 \left[\theta(t) - \theta_u(t)\right]^2 + w_2 \left(\left[x(t) - x_u(t)\right]^2 + \left[y(t) - y_u(t)\right]^2 \right) \right] dt \quad (7.4)$$

where the weights, $w_1, w_2 \in [0, 1]$ with $w_1 + w_2 = 1$, determine the relative importance of the translational and orientation errors.

As stated above, we are concerned with the situation in which the unicycle is rotating, but has no linear motion. In this section we develop an open-loop optimal control scheme to handle this case. In particular, the unicycle trajectory that the car must follow, shown in Figure 7.1, starts at the origin and includes a sharp turn. This trajectory is generated by the unicycle inputs

$$v_1(t) = \begin{cases} v_s & t \in [0, t_s) \\ 0 & t \in [t_s, t_s + t_t) \\ v_s & t \in [t_s + t_t, 2t_s + t_t] \end{cases} \quad (7.5)$$

$$v_2(t) = \begin{cases} 0 & t \in [0, t_s) \\ \omega & t \in [t_s, t_s + t_t) \\ 0 & t \in [t_s + t_t, 2t_s + t_t] \end{cases} \quad (7.6)$$

where v_s and t_s determine the length of the straight portion and ω and t_t determine the angle of the turn.

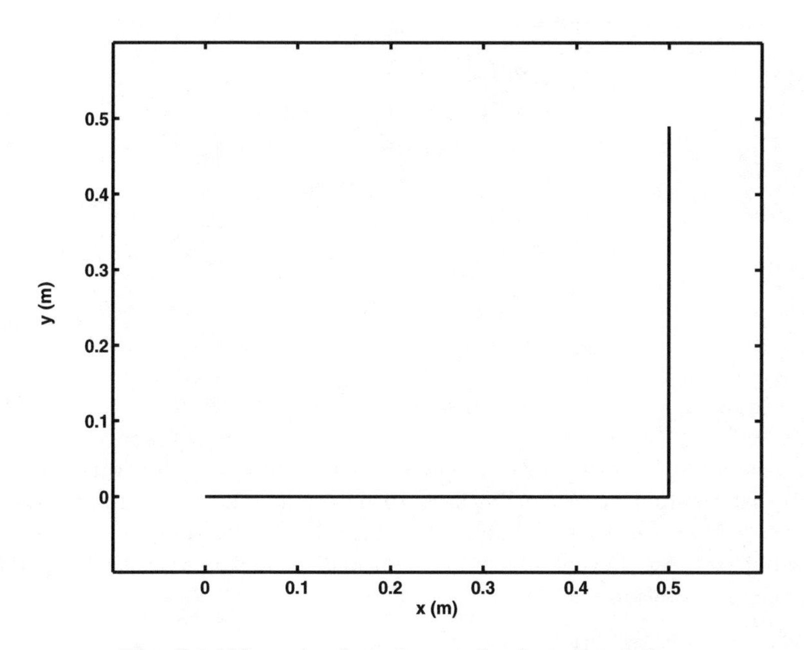

Fig. 7.1. The unicycle trajectory for the car to follow

Given the cost function (7.4), we can define the Hamiltonian as

$$\begin{aligned} H =& \frac{1}{2}w_1 \left[\theta(t) - \theta_u(t)\right]^2 + \frac{1}{2}w_2 \left(\left[x(t) - x_u(t)\right]^2 + \left[y(t) - y_u(t)\right]^2\right) \\ &+ p_1(t)v_1(t)\cos\theta(t) + p_2(t)v_1(t)\sin\theta(t) \\ &+ p_3(t)v_1(t)\frac{\tan\phi(t)}{l} + p_4(t)v_2(t). \end{aligned} \tag{7.7}$$

If we denote $[x \ y \ \theta \ \phi]^T$ by X and $[p_1 \ p_2 \ p_3 \ p_4]^T$ by p, then the necessary conditions for optimal control are given by [26]

$$\dot{X}^*(t) = \frac{\partial H}{\partial p} = \begin{bmatrix} v_1^*(t)\cos\theta^*(t) \\ v_1^*(t)\sin\theta^*(t) \\ v_1^*(t)\frac{\tan\phi^*(t)}{l} \\ v_2^*(t) \end{bmatrix} \tag{7.8}$$

$$\dot{p}^*(t) = -\frac{\partial H}{\partial X} = -\begin{bmatrix} w_2\left[x^*(t) - x_u(t)\right] \\ w_2\left[y^*(t) - y_u(t)\right] \\ w_1\left[\theta^*(t) - \theta_u(t)\right]p_1^*(t)v_1^*(t)\sin\theta^*(t) \\ +p_2^*(t)v_1^*(t)\cos\theta^*(t)p_3^*(t)v_1^*(t)\frac{\sec^2\phi^*(t)}{l} \end{bmatrix} \tag{7.9}$$

$$0 = \frac{\partial H}{\partial u} = \begin{bmatrix} p_1^*(t)\cos\theta^*(t) + p_2^*(t)\sin\theta^*(t) + p_3^*(t)\frac{\tan\phi^*(t)}{l} \\ p_4^*(t) \end{bmatrix} \tag{7.10}$$

where the * denotes the optimal values.

The boundary conditions are

$$x^*(0) = x_u(0) \tag{7.11}$$
$$y^*(0) = y_u(0) \tag{7.12}$$
$$\theta^*(0) = \theta_u(0) \tag{7.13}$$
$$\phi^*(0) = \tan^{-1} \frac{l\dot{\theta}_u(0)}{v_1(0)} \tag{7.14}$$
$$x^*(t_f) = x_u(t_f) \tag{7.15}$$
$$y^*(t_f) = y_u(t_f) \tag{7.16}$$
$$\theta^*(t_f) = \theta_u(t_f) \tag{7.17}$$
$$\phi^*(t_f) = \tan^{-1} \frac{l\dot{\theta}_u(t_f)}{v_1(t_f)}. \tag{7.18}$$

These boundary conditions were chosen so that before and after this maneuver, the transformation (7.3) can be used to determine the car's inputs from the unicycle's. However, these equations are difficult to solve simultaneously, so the following steepest descent algorithm, adapted from [26], was used to find an approximate solution.

1. Choose an initial $v_1(t)$ and $v_2(t)$.
2. Using the initial conditions (7.11)-(7.14), integrate the car dynamics (3.10) to obtain the car's trajectory $x(t)$, $y(t)$, $\theta(t)$, and $\phi(t)$ for $t \in [0, t_f]$.
3. Using the boundary condition $p(t_f) = 0$ and (7.9), backward integrate to obtain the p_i's.
4. Calculate the cost using (7.4).
5. If the cost is higher than some predetermined threshold, update the inputs according to

$$v_1^{(i+1)} = v_1^{(i)} - \tau \frac{\partial H}{\partial v_1} \tag{7.19}$$
$$v_2^{(i+1)} = v_2^{(i)} - \tau \frac{\partial H}{\partial v_2} \tag{7.20}$$

   where τ is the step size constant.

Steps 2-5 are repeated until the value of the cost function is sufficiently low.

This algorithm finds the costates (p_i's) by forcing their final values to zero. However, the boundary condition on the final states, $x(t_f), y(t_f), \theta(t_f), \phi(t_f)$, is not enforced. As a result, the car's final state is allowed to deviate from the unicycle's final state.

The success of this algorithm depends on the initial input. Two initial inputs were used that were considered to be near the optimum. One input had the car follow an inscribed circle as shown in Figure 7.2, deviating from the unicycle trajectory before the turning point was reached. The other input caused the car to follow the unicycle trajectory until the turning point at which time the car followed two sequential arcs before joining the unicycle trajectory after the turn. An example of this path is shown in Figure 7.3.

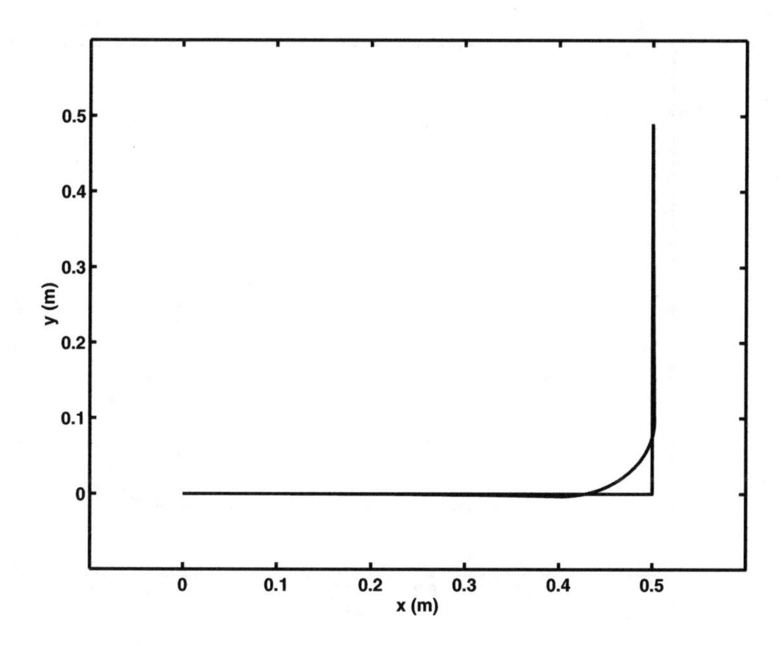

Fig. 7.2. The car follows an inscribed circle based on its maximum steering angle

The first trajectory was created by determining the two points at which the inscribed circle touched the unicycle trajectory. This is shown in Figure 7.4 for the general case of any angle α. The radius of the circle depends on the car's maximum steering angle (in radians) and the wheelbase length:

$$r = \frac{l}{\tan \phi_{max}}. \tag{7.21}$$

The distance d_1 from where the circle touches the unicycle trajectory to the turning point is given by

$$d_1 = \frac{r}{\tan \frac{\alpha}{2}}. \tag{7.22}$$

The car follows the unicycle trajectory until it reaches the first intersection point, whose x-coordinate is given by

$$x_c = v_s t_s - d_1. \tag{7.23}$$

Then the car must follow the arc defined by β for the same amount of time that the unicycle traveled distance d_1, turned, and traveled distance d_1 again. Thus the car's linear velocity during this time is

$$v_{1_{turn}} = \frac{s}{t_{arc}} \tag{7.24}$$

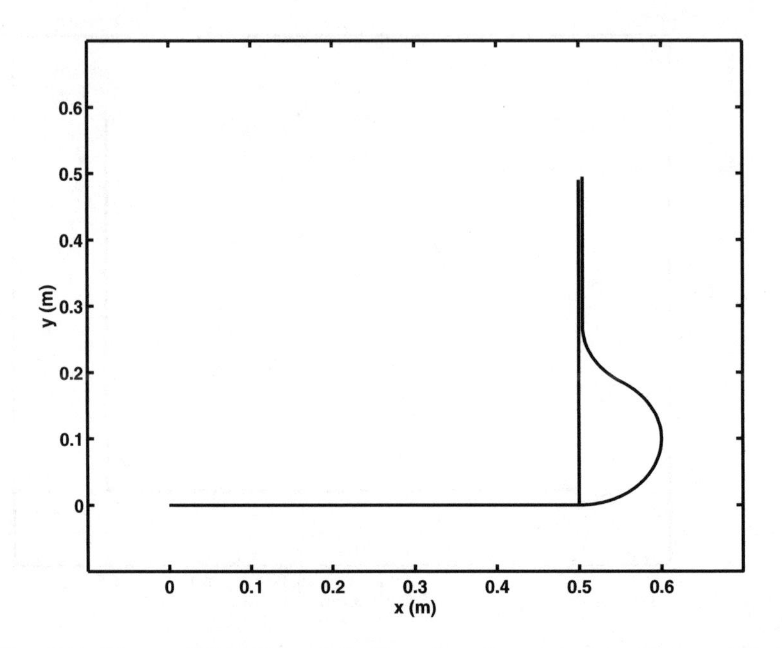

Fig. 7.3. The car follows two circles based on its maximum steering angle

where

$$s = r\beta \tag{7.25}$$

$$\beta = \pi - \alpha \tag{7.26}$$

$$t_{arc} = \frac{2d_1}{v_s} + t_t \tag{7.27}$$

The car control inputs that generate this trajectory are

$$v_1(t) = \begin{cases} v_s & t \in [0, t_c) \\ v_{1_{turn}} & t \in [t_c, t_c + t_{arc}) \\ v_s & t \in [t_c + t_{arc}, t_f] \end{cases} \tag{7.28}$$

$$v_2(t) = \begin{cases} 0 & t \in [0, t_c) \\ \frac{v_{2max}}{2} & t = t_c \\ 0 & t \in (t_c, t_c + t_{arc}) \\ \frac{-v_{2max}}{2} & t = t_c + t_{arc} \\ 0 & t \in (t_c + t_{arc}, t_f] \end{cases} \tag{7.29}$$

where $t_c = \frac{x_c}{v_s}$ and $\frac{v_{2max}}{2}$ is the input that causes the wheels to instantaneously turn through $\frac{\phi_{max}}{2}$.

For the second initial trajectory, the car followed the unicycle until the turning point. Then the car made two turns, following two arcs determined by the car's

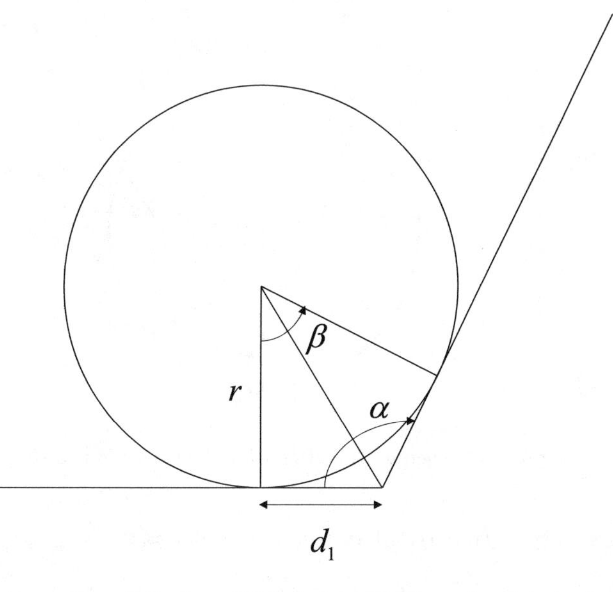

Fig. 7.4. The intersection of the inscribed circle with the unicycle's trajectory

maximum steering angle. The construction of the two arcs is shown in Figure 7.5. The distance d_2 between the intersection of each circle with the unicycle's trajectory is given by

$$d_2 = r \sin\alpha + \sqrt{3 - \cos^2\alpha + 2\cos\alpha}. \tag{7.30}$$

The angles, β_1 and β_2, that define the two arcs, s_1 and s_2, are

$$\beta_1 = \pi - \alpha + \tan^{-1}\left[\frac{\sqrt{3 - \cos^2\alpha + 2\cos\alpha}}{(1-\cos\alpha)(\cos\alpha)} - \frac{\sqrt{3 - \cos^2\alpha + 2\cos\alpha}}{\cos\alpha}\right] \tag{7.31}$$

$$\beta_2 = \tan^{-1}\left[\frac{\sqrt{3 - \cos^2\alpha + 2\cos\alpha}}{1-\cos\alpha}\right]. \tag{7.32}$$

The car follows the unicycle trajectory until the turn, whose x-coordinate is given by

$$x_s = v_s t_s. \tag{7.33}$$

As before, the car must travel the distance $s_1 + s_2$ in the same time it takes the unicycle to complete the turn and travel distance d_2. The car's velocity during this time is

$$v_{1_{turns}} = \frac{s}{t_{turns}} \tag{7.34}$$

where $s = r(\beta_1 + \beta_2)$ and $t_{turns} = t_t + \frac{d_2}{v_s}$.

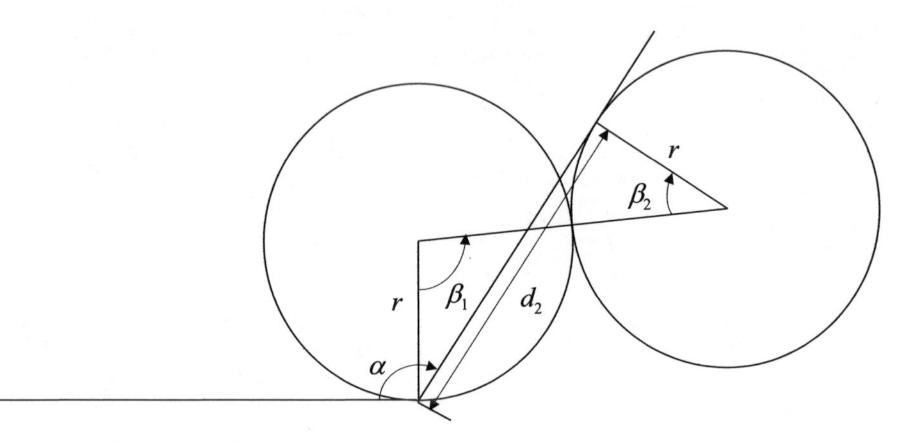

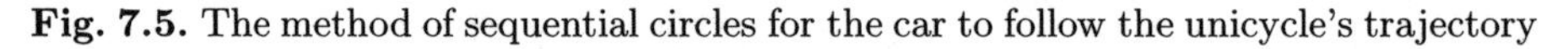

Fig. 7.5. The method of sequential circles for the car to follow the unicycle's trajectory

The car inputs that result in this trajectory are

$$v_1 = \begin{cases} v_s & t \in [0, t_s) \\ v_{1_{turns}} & t \in [t_s, t_s + t_{turns}) \\ v_s & t \in [t_s + t_{turns}, t_f] \end{cases} \tag{7.35}$$

$$v_2 = \begin{cases} 0 & t \in [0, t_s) \\ \frac{v_{2_{max}}}{2} & t = t_s \\ 0 & t \in (t_s, t_s + t_1) \\ -v_{2_{max}} & t = t_s + t_1 \\ 0 & t \in (t_s + t_1, t_s + t_1 + t_2) \\ \frac{v_{2_{max}}}{2} & t = t_s + t_1 + t_2 \\ 0 & t \in (t_s + t_1 + t_2, t_f] \end{cases} \tag{7.36}$$

where $t_1 = \frac{\beta_1}{\beta_1 + \beta_2} t_{turns}$ and $t_2 = \frac{\beta_2}{\beta_1 + \beta_2} t_{turns}$.

7.3 Simulation

The optimal control algorithm described in the previous section was simulated in MATLAB. First, the starting trajectories were generated and the cost function was calculated for various weights and maximum steering angles. The results are shown in Table 7.1.

The MATLAB simulation was run for three different initial inputs: the two described in the previous section and one that causes the car to drive on a straight trajectory

$$v_1(t) = v(t) \tag{7.37}$$

$$v_2(t) = 0. \tag{7.38}$$

Table 7.1. The value of the cost function J for various weights and maximum steering angles

ϕ_{max}	$w_1 = 0, w_2 = 1$	$w_1 = w_2 = 0.5$	$w_1 = 1, w_2 = 0$
$\frac{\pi}{3}$	0.0774	0.1006	0.1238
$\frac{\pi}{4}$	0.2100	0.2771	0.3441
$\frac{\pi}{5}$	0.4022	0.5070	0.6119
$\frac{\pi}{6}$	0.5837	0.7507	0.9178

Table 7.2. The value of the cost function J for the different initial inputs

Initial Input	$w_1 = 0, w_2 = 1$	$w_1 = w_2 = 0.5$	$w_1 = 1, w_2 = 0$
Inscribed Circle	0.1329	0.1089	0.0127
Sequential Circles	0.4383	0.3419	0.0253
Straight Trajectory	0.5125	1.2395	0.0767

Table 7.3. The number of iterations for the different initial inputs

Initial Input	$w_1 = 0, w_2 = 1$	$w_1 = w_2 = 0.5$	$w_1 = 1, w_2 = 0$
Inscribed Circle	307	182	142
Sequential Circles	772	427	294
Straight Trajectory	1001	1001	612

In all cases the car's initial conditions were $x(0) = x_u(0)$, $y(0) = y_u(0)$, $\theta(0) = \theta_u(0)$, and $\phi(0) = \tan^{-1}\left(\frac{l\omega(0)}{v(0)}\right)$. The car constants l and ϕ_{max} were set to 0.1 and $\frac{\pi}{4}$ respectively. The step size τ was 0.01, the sampling time T was 0.01, and the maximum value for v_2 was 100000. The simulation was run for 1000 iterations or until the change in the cost function J was less than 0.0001, whichever occurred first. The results are summarized in Tables 7.2 and 7.3.

In all three cases, the lowest cost and the fastest convergence were obtained with $w_1 = 1$ and $w_2 = 0$, i.e., when only the orientation error was considered. For the initial straight trajectory when the x, y-error was included in the cost function, the algorithm did not converge in 1000 steps. The inscribed circle converged to its result the fastest and had the overall lowest cost. The final trajectory and inputs for this case are shown in Figures 7.6 and 7.7. The final position of the car does not match that of the unicycle exactly. This was the case for all of the trajectories resulting from the optimal control algorithm. Because the final costates (p_i's) were forced to zero, the boundary conditions for the car's final state (7.15)-(7.18) were not enforced.

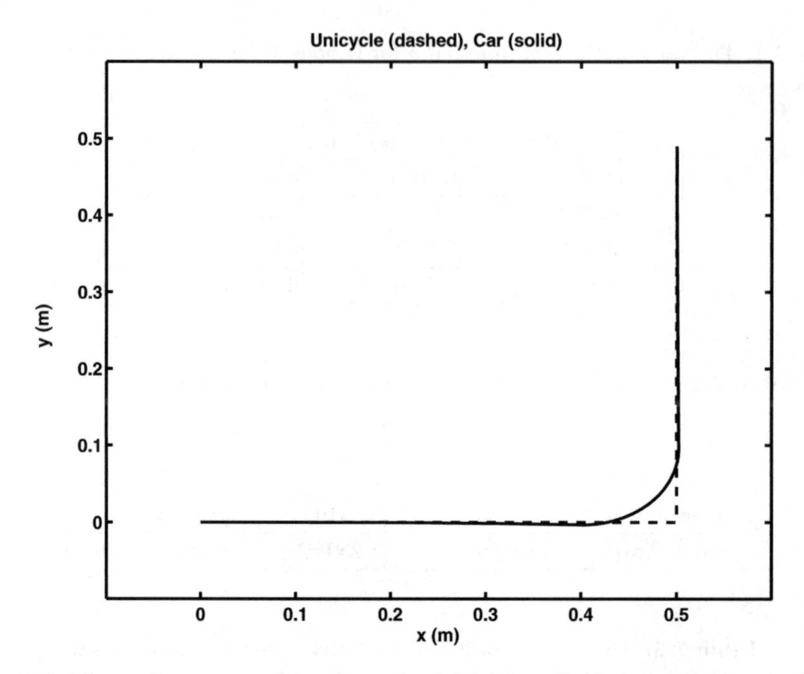

Fig. 7.6. The trajectory resulting from the initial inscribed circle with $w_1 = 1$ and $w_2 = 0$

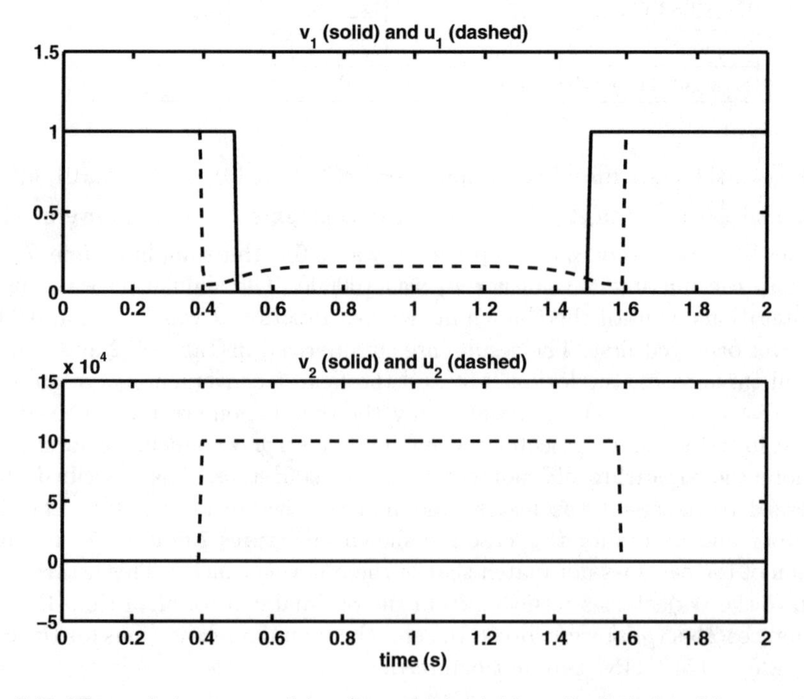

Fig. 7.7. The car's inputs resulting from the initial inscribed circle with $w_1 = 1$ and $w_2 = 0$

8 Uncertainty Propagation in Abstracted Systems

In this chapter, it is shown that given a system and its abstraction, the evolution of uncertain initial conditions in the original system is, in some sense, matched by the evolution of the uncertainty in the abstracted system. In other words, it is shown that the concept of Φ-related vector fields extends to the case of stochastic initial conditions where the probability density function (pdf) for the initial conditions is known. In the deterministic case, the Φ mapping commutes with the system dynamics. In this chapter, it is shown that in the case of stochastic initial conditions, the induced mapping, Φ_{pdf}, commutes with the evolution of the pdf according to the Liouville equation. It is also shown that a control system abstraction can capture the time evolution of the uncertainty in the original system by an appropriate choice of control input. Application of the convservation law results in a partial differential equation known as the Liouville equation, for which a closed form solution is known. The solution provides the time evolution of the initial pdf which can be followed by the abstracted system.

The propagation of uncertain initial conditions in dynamical systems has been studied in different contexts using various approaches. Monte Carlo methods [50] can be used to track the trajectories of a specific realization of the initial condition to generate the probability distributions after any time. However, this method is computationally expensive since simulations must be run millions of times to achieve a distribution. Polynomical chaos expansion is a method that requires less computation than Monte Carlo, but results in an approximation of the distribution. Using polynomial chaos, the random variable is expressed as a summation of weighted basis polynomials. These basis polynomials can be chosen to achieve the fastest convergence based on the type of uncertainty [61]. This method has been used to study stability of systems with uncertain parameters [18]. Another method of studying the propagation of uncertain initial conditions has been to use conservation concepts to find an exact expression (via the Liouville equation) for the time evolution of the uncertainty [11].

Recently, the notions of abstraction and stochastic behavior have been combined. These ideas have been studied in the context of stochastic hybrid systems in [22] and [46]. In [59], equivalence is defined between systems where dynamics themselves contain stochastic elements rather than the states or parameters.

P. Mellodge & P. Kachroo: Model Abstraction in Dynamical Systems, LNCIS 379, pp. 97–110, 2008.
springerlink.com

In this chapter, it is assumed that the system dynamics are deterministic while only the initial conditions are stochastic. Furthermore, it is assumed that the pdf of the initial condition is known and that this pdf evolves in time according to the Liouville equation. Our goal is to show that the solution to the Liouville equation and the induced mapping, denoted by Φ_{pdf}, commute. This result is similar to that for for the deterministic case. The Φ-relationship for systems with known initial conditions is shown in Figure 8.1 and for systems with stochastic initial conditions is shown in Figure 8.2.

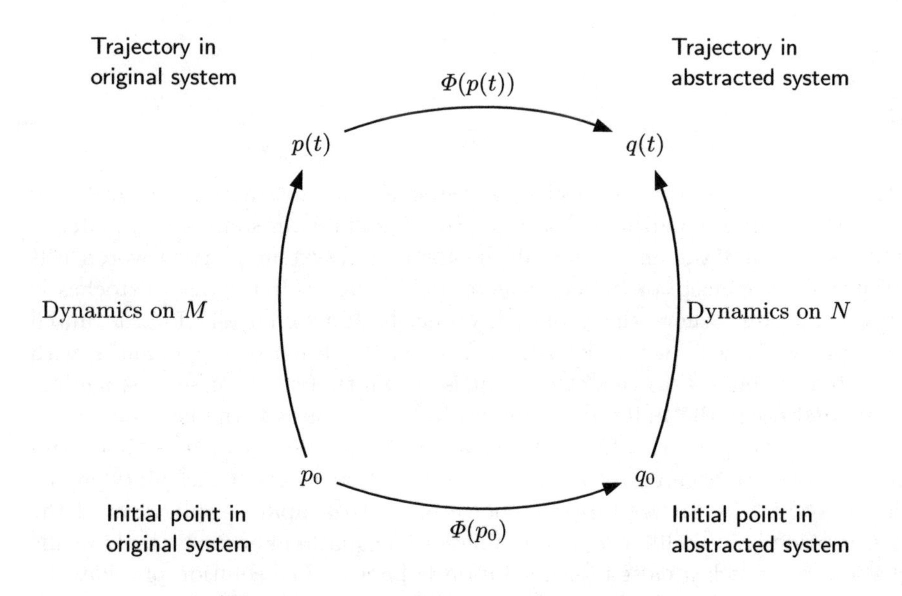

Fig. 8.1. The relationship between Φ-related systems with known initial conditions

It is intuitively clear that the commutative relationship is true if one considers the Monte Carlo approach to the problem. Suppose the original system has dynamics defined on manifold M and it is Φ-related to a system defined on manifold N. Choosing initial conditions in the original system according to its distribution, by following the dynamics on M for some time t and then Φ-mapping the endpoint of the trajectory to the abstracted system, produces $q \in N$ at time t. If the initial point on M were first mapped to N and the dynamics on N followed, the same q at time t would result. This behavior is due to the Φ-relatedness of the two systems in the deterministic setting. Repeating this procedure many times by choosing initial conditions that follow its distribution, the resulting distribution for q at time t is independent of which operation is performed first. In this chapter, it is shown that this intuitive notion is correct by using the Liouville equation as a way to propagate the distribution of the initial condition though the dynamics of the system.

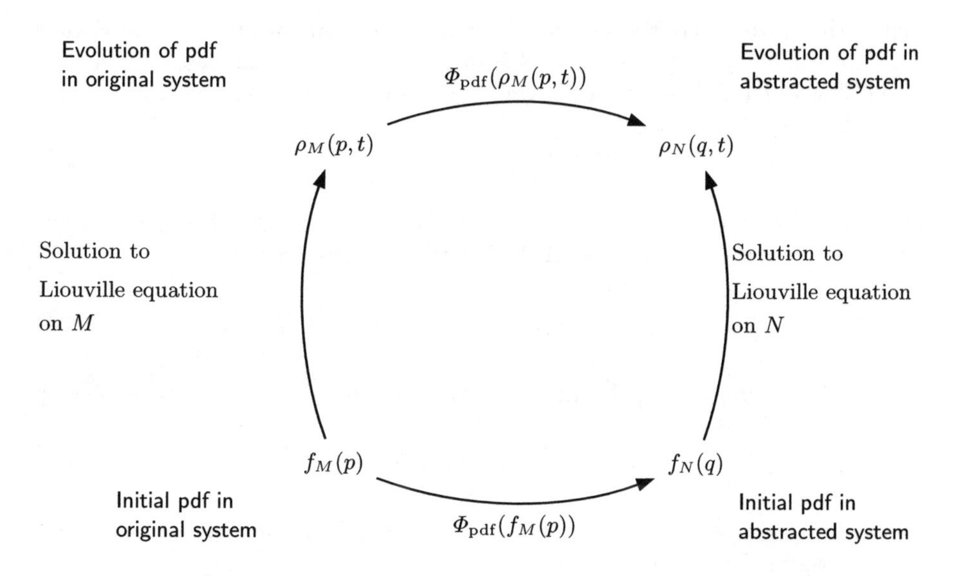

Fig. 8.2. The relationship between Φ-related systems with uncertain initial conditions

8.1 The Liouville Equation and Its Solution

In this section reviews the Liouville equation and its solution as found in [12]. It should be noted that the notions of systems, probability distributions, and Φ-relatedness are coordinate-free. That is, they are defined with respect to points on a manifold, independent of the local description of those points. Throughout this paper, the derivations are given using local representations of the systems and probability distributions. The results, however, remain valid for any choice of coordinates and are thus independent of these representations.

8.1.1 Derivation of the Liouville Equation

Consider the autonomous system given by

$$\dot{x} = F(x) \tag{8.1}$$

where $x = (x_1, \ldots, x_n)$.

Suppose that the initial condition, $x(0)$, is unknown, but its statisical properties are known through its pdf, $f(x)$. Denote the time evolution of the pdf by $\rho(x,t)$, where $\rho(x,0) = f(x)$. Since $\rho(x,t)$ must be a pdf, it is known that $\rho(x,t) \geq 0$ for all x and t and that its integral over the entire state space must be unity for every t. Viewing this distribution as a mass to be conserved, the Eulerian approach from fluid dynamics can be employed to derive the continuity equation for $\rho(x,t)$ given the dynamics $F(x)$. That is, for an infinitesimal area of the state space, the net flow leaving the volume must equal the change in volume under $\rho(x,t)$ (see Figure 8.3 for the one dimensional case). If the flow in and flow out are denoted by q_{in} and q_{out}, respectively, then

$$\frac{\partial \rho(x,t)}{\partial t}\Delta x = q_{in} - q_{out}. \tag{8.2}$$

The flow entering the left side of Δx is $\rho(x,t)F(x)$ and the flow leaving the right side of Δx is $\rho(x+\Delta x,t)F(x+\Delta x)$. Substituting into (8.2) yields

$$\frac{\partial \rho(x,t)}{\partial t}\Delta x = \rho(x,t)F(x) - \rho(x+\Delta x,t)F(x+\Delta x). \tag{8.3}$$

Dividing by Δx and taking the limit gives the Liouville equation in one dimension

$$\frac{\partial \rho(x,t)}{\partial t} = -\frac{\partial[\rho(x,t)F(x)]}{\partial x}. \tag{8.4}$$

For higher dimensions, the net flow leaving a volume is given by the divergence of the flow, resulting in the general equation,

$$\frac{\partial \rho(x,t)}{\partial t} + \sum_{i=1}^{n} \frac{\partial}{\partial x_i}\left[\rho(x,t)F_i(x)\right] = 0 \tag{8.5}$$

where $F_i(x)$ is the ith component of $F(x)$.

A more complete discussion of this derivation can be found in [12].

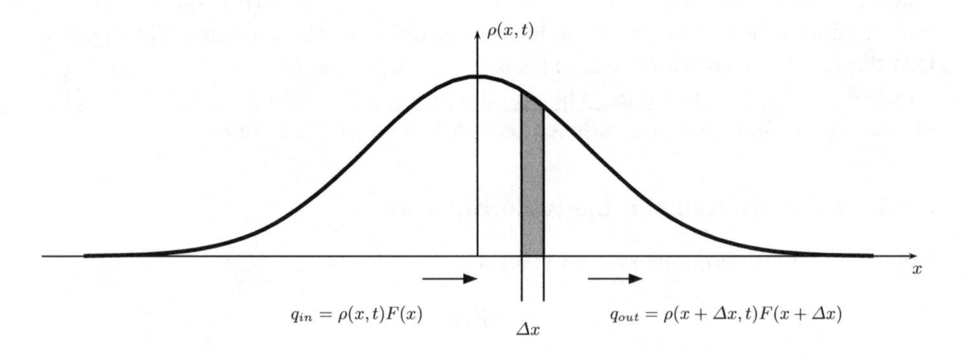

Fig. 8.3. The area under $\rho(x,t)$ must be conserved. The change in shaded area is equal to the difference between the flow into and out of the area Δx.

8.1.2 Solution to the Liouville Equation

In this section, the Liouville equation is solved to provide an explicit expression for $\rho(x,t)$. First, notation similar to that used in [11] is introduced.

For the system (8.1), let the initial condition be denoted by Ξ. That is, $\Xi = x(0)$. Assuming that the solution to (8.1) exists and is unique, then there is a unique Ξ corresponding to every $x(t)$. Thus,

$$\Xi = \Xi(x,t). \tag{8.6}$$

Expanding (8.5) yields

$$\frac{\partial \rho(x,t)}{\partial t} + \sum_{i=1}^{n} F_i(x) \frac{\partial \rho(x,t)}{\partial x_i} + \rho(x,t) \sum_{i=1}^{n} \frac{\partial F_i(x)}{\partial x_i} = 0. \tag{8.7}$$

The first two terms on the left side of (8.7) express the full derivative of $\rho(x,t)$ with respect to t. Using the definition,

$$\psi(x) = \sum_{i=1}^{n} \frac{\partial F_i(x)}{\partial x_i} \tag{8.8}$$

then (8.7) can be rewritten as

$$\frac{d\rho(x,t)}{dt} = -\rho(x,t)\psi(x). \tag{8.9}$$

Separating variables and integrating gives

$$\int_{\rho(x(0),0)}^{\rho(x,t)} \frac{1}{\rho(x,\tau)} \, d\rho(x,\tau) = -\int_0^t \psi(x(\tau)) \, d\tau. \tag{8.10}$$

It is important to note that, although x is a function of t in terms of the system dynamics, x and t are independent when evaluating $\rho(x,t)$. Using the fact that $\Xi(x,t)$ is the initial condition of the system associated with x and t, and that $\rho(x(0),0) = f(\Xi(x,t))$, the solution to the Liouville equation is

$$\rho(x,t) = f(\Xi(x,t)) \exp\left[-\int_0^t \psi(\hat{x}(\tau)) \, d\tau\right] \tag{8.11}$$

where $\hat{x}(\tau)$ denotes the trajectory starting at $\Xi(x,t)$ at time zero and ending at x at time t.

This solution leads to some results relating the transformation from x to Ξ with the divergence of the vector field $\psi(x)$. Since the area under the pdf is conserved, then

$$\rho(x,t) \, dx_1 \cdots dx_n = f(\Xi(x,t)) \, d\Xi_1 \cdots d\Xi_n. \tag{8.12}$$

By the change of variables theorem, the areas $dx_1 \cdots dx_n$ and $d\Xi_1 \cdots d\Xi_n$ are related by the determinant of the Jacobian of the transformation from x to Ξ, i.e.,

$$dx_1 \cdots dx_n = \left| \frac{\partial x}{\partial \Xi} \right| d\Xi_1 \cdots d\Xi_n. \tag{8.13}$$

Substituting this expression into (8.12) gives

$$\frac{\rho(x,t)}{f(\Xi(x,t))} = \left|\frac{\partial x}{\partial \Xi}\right|^{-1}. \tag{8.14}$$

Then by (8.11),

$$\exp\left[\int_0^t \psi(\hat{x}(\tau))\, d\tau\right] = \left|\frac{\partial x}{\partial \Xi}\right|. \tag{8.15}$$

8.2 Definition of the Φ_{pdf} Mapping

Before extending the solution of the Liouville equation to Φ-related systems, the definition of Φ_{pdf} is provided. This is the mapping of the pdf in the original system to the abstracted one, as shown in Figure 8.2.

Assume that the systems

$$\dot{x} = F(x) \tag{8.16}$$
$$\dot{y} = G(y) \tag{8.17}$$

are Φ-related systems such that

$$y = \Phi(x) \tag{8.18}$$

and

$$G(y) = T\Phi(F(x)) \tag{8.19}$$

with $x = (x_1, \ldots, x_n) \in \mathbb{R}^n$, $y = (y_1, \ldots, y_m) \in \mathbb{R}^m$ and $m \leq n$. The system in (8.16) is the original system and its abstraction is given by (8.17). Since Φ provides a static map from $\mathbb{R}^n$ to $\mathbb{R}^m$, a pdf in for the initial condition of the original system can be mapped to one for the initial condition of the abstracted system. Thus, given a pdf in $\mathbb{R}^n$, $f_X(x)$, the corresponding pdf in $\mathbb{R}^m$ is

$$f_Y(y) = \Phi_{\mathrm{pdf}}(\mathrm{f_X(x)}) \tag{8.20}$$
$$= \frac{d^m}{dy_1 \cdots dy_m} \int_{D(y)} f_X(x) dx_1 \cdots dx_n \tag{8.21}$$

where

$$D(y) = \{x : \Phi_1(x) \leq y_1, \ldots, \Phi_m(x) \leq y_m\}. \tag{8.22}$$

As noted in the previous section, the notions of systems, probability distributions, and Φ-relatedness are coordinate free. Although described above in terms of local coordinates $(x_1, \ldots, x_n)$ and $(y_1, \ldots, y_m)$, Φ_{pdf} is a global mapping between probability distributions on two different manifolds. Given a pdf defined on the original system's state manifold, Φ_{pdf} gives the corresponding pdf on the abstracted state manifold. The mapping between pdf's is induced by Φ, the mapping between points.

8.3 Example of Φ-Related Linear Systems

In this section, an example is given to illustrate the steps in Figure 8.2. Although this example uses a very simple system, it helps to determine the derivation steps for the more general case.

Consider the system

$$\begin{aligned}\dot{x}_1 &= ax_1\\ \dot{x}_2 &= bx_2\end{aligned} \tag{8.23}$$

with $x_1, x_2 \in \mathbb{R}$. Let $\Phi : \mathbb{R}^2 \to \mathbb{R}$, with $\Phi(x_1, x_2) = x_1$. Then

$$\dot{x}_1 = ax_1. \tag{8.24}$$

An abstraction of (8.23) is given by

$$\dot{y} = ay \tag{8.25}$$

with $y \in \mathbb{R}$. The systems (8.23) and (8.25) are Φ-related since (8.25) captures the behavior in (8.24). Note that the divergences of vector fields in each system are constant and are given by

$$\psi_X(x) = a + b \tag{8.26}$$

$$\psi_Y(y) = a. \tag{8.27}$$

Let $\Xi_x = (\Xi_{x_1}, \Xi_{x_2})$ be the initial condition associated with the point (x_1, x_2) at time t and Ξ_y be that associated with y at time t.

Given the initial pdf for the original system, $f_X(x_1, x_2)$, the corresponding initial pdf for the abstracted system is

$$f_Y(y) = \frac{d}{dy} \int\int_{\{(x_1,x_2):\Phi(x_1,x_2)\leq y\}} f_X(x_1, x_2)\, dx_1\, dx_2 \tag{8.28}$$

$$= \frac{d}{dy} \int_{-\infty}^{\infty} \int_{-\infty}^{y} f_X(x_1, x_2)\, dx_1\, dx_2. \tag{8.29}$$

The time evolution of (8.29) is given by the solution to the Liouville equation

$$\tilde{\rho}_Y(y, t) = f_Y(\Xi_y(y, t)) \exp\left[-\int_0^t a\, d\tau\right] \tag{8.30}$$

$$= \left(\frac{d}{d\Xi_y} \int_{-\infty}^{\infty} \int_{-\infty}^{\Xi_y(y,t)} f_X(x_1, x_2)\, dx_1\, dx_2\right) e^{-at} \tag{8.31}$$

$$= \left(\int_{-\infty}^{\infty} f_X(\Xi_y(y, t), x_2)\, dx_2\right) e^{-at}. \tag{8.32}$$

The step from (8.31) to (8.32) is an application of Leibnitz's rule. The time evolution of $f_X(x_1, x_2)$ is given by the solution to the Liouville equation

$$\rho_X(x_1, x_2, t) = f_X(\Xi_{x_1}(x_1, x_2, t), \Xi_{x_2}(x_1, x_2, t)) \exp\left[-\int_0^t (a + b)\, d\tau\right] \tag{8.33}$$

$$= f_X(\Xi_{x_1}(x_1, x_2, t), \Xi_{x_2}(x_1, x_2, t))\, e^{-(a+b)t}. \tag{8.34}$$

Mapping (8.34) through Φ_{pdf} gives

$$\rho_Y(y,t) = \frac{d}{dy}\int_{-\infty}^{\infty}\int_{-\infty}^{y} \rho_X(x_1,x_2,t)\,dx_1\,dx_2 \tag{8.35}$$

$$= \frac{d}{dy}\int_{-\infty}^{\infty}\int_{-\infty}^{y} f_X(\Xi_{x_1}(x_1,x_2,t),\Xi_{x_2}(x_1,x_2,t))\,e^{-(a+b)t}\,dx_1\,dx_2 \tag{8.36}$$

$$= \frac{d}{dy}\int_{-\infty}^{\infty}\int_{-\infty}^{\Xi_y(y,t)} f_X(\Xi_{x_1}(x_1,x_2,t),\Xi_{x_2}(x_1,x_2,t))$$

$$e^{-(a+b)t}\left|\frac{\partial x}{\partial \Xi_x}\right| d\Xi_{x_1}\,d\Xi_{x_2}. \tag{8.37}$$

The closed form solution for the original system is given by

$$\begin{bmatrix} x_1(t) \\ x_2(t) \end{bmatrix} = \begin{bmatrix} e^{at} & 0 \\ 0 & e^{bt} \end{bmatrix}\begin{bmatrix} \Xi_{x_1} \\ \Xi_{x_2} \end{bmatrix}. \tag{8.38}$$

So the determinant of the Jacobian is

$$\left|\frac{\partial x}{\partial \Xi_x}\right| = e^{(a+b)t}. \tag{8.39}$$

Substituting for the Jacobian and applying Leibnitz's rule gives

$$\rho_Y(y,t) = \frac{d}{dy}\int_{-\infty}^{\infty}\int_{-\infty}^{\Xi_y(y,t)} f_X(\Xi_{x_1}(x_1,x_2,t),\Xi_{x_2}(x_1,x_2,t))\,d\Xi_{x_1}\,d\Xi_{x_2} \tag{8.40}$$

$$= \int_{-\infty}^{\infty} f_X(\Xi_y(y,t),\Xi_{x_2}(x_1,x_2,t))\frac{\partial \Xi_y(y,t)}{\partial y}\,d\Xi_{x_2}. \tag{8.41}$$

From the closed form solution of the abstracted system, the initial condition can be expressed as

$$\Xi_y(y,t) = e^{-at}y \tag{8.42}$$

so that

$$\frac{\partial \Xi_y(y,t)}{\partial y} = e^{-at}. \tag{8.43}$$

The time evolution of the pdf is then

$$\rho_Y(y,t) = \left(\int_{-\infty}^{\infty} f_X(\Xi_y(y,t),\Xi_{x_2}(x_1,x_2,t))\,d\Xi_{x_2}\right)e^{-at}. \tag{8.44}$$

Thus, the expression for $\tilde{\rho}_Y(y,t)$ given by (8.32) is equivalent to (8.44). This shows that the solution to the Liouville equation commutes with the mapping Φ_{pdf}. Figure 8.4 shows $\rho_Y(y,t)$ for $a=-1$ and Gaussian initial pdf with $\mu=15$ and $\sigma=1$. It can be seen in the figure that as time progresses, the mean and variance approach zero since all realizations of the initial condition go to zero. In the limit, the distribution becomes the unit impulse function, reflecting the fact all realizations go to zero and the state is known with certainty.

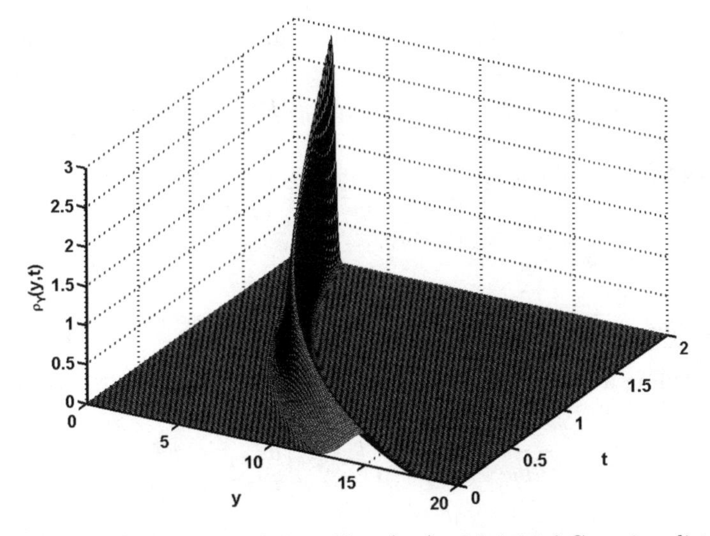

Fig. 8.4. The time evolution of the pdf $\rho_Y(y,t)$ with initial Gaussian distribution

8.4 Autonomous Φ-Related Systems

Consider the general case of Φ-related systems, described locally as

$$\dot{x} = F(x) \tag{8.45}$$

$$\dot{y} = G(y) \tag{8.46}$$

with $x \in \mathbb{R}^n$, $y \in \mathbb{R}^m$, and

$$y = \Phi(x) \tag{8.47}$$

$$G(y) = T\Phi(F(x)) \tag{8.48}$$

where $T\Phi$ is the induced mapping between vector fields. As before, the initial condition associated with x at time t is $\Xi_x = (\Xi_{x_1}, \ldots, \Xi_{x_n})$ and that associated with y at time t is $\Xi_y = (\Xi_{y_1}, \ldots, \Xi_{y_m})$. Let $f_X(x)$ be the initial pdf associated with x. Then $f_Y(y)$ is given by (8.21)

$$f_Y(y) = \frac{d^m}{dy_1 \cdots dy_m} \int_{D(y)} f_X(x) dx_1 \cdots dx_n. \tag{8.49}$$

The solution to the Liouville equation starting with $f_Y(y)$ is

$$\tilde{\rho}_Y(y,t) = f_Y(\Xi_y(y,t)) \exp\left[-\int_0^t \psi_Y(\hat{y}(\tau))d\tau\right] \tag{8.50}$$

$$= \left(\frac{d^m}{d\Xi_{y_1} \cdots d\Xi_{y_m}} \int_{D(\Xi_y(y,t))} f_X(x) dx_1 \cdots dx_n\right)$$

$$\exp\left[-\int_0^t \psi_Y(\hat{y}(\tau))d\tau\right] \tag{8.51}$$

$$= \left(\frac{d^m}{d\Xi_{y_1} \cdots d\Xi_{y_m}} \int_{D(\Xi_y(y,t))} f_X(x) dx_1 \cdots dx_n\right) \left|\frac{\partial y}{\partial \Xi_y}\right|^{-1}. \tag{8.52}$$

Now, the solution to the Liouville equation in the original system is

$$\rho_X(x,t) = f_X(\Xi_x(x,t)) \exp\left[-\int_0^t \psi_X(\hat{x}(\tau))d\tau\right]. \tag{8.53}$$

Mapping (8.53) through Φ_{pdf} gives

$$\rho_Y(y,t) = \frac{d^m}{dy_1 \cdots dy_m} \int_{D(y)} \rho_X(x,t) dx_1 \cdots dx_n \tag{8.54}$$

$$= \frac{d^m}{dy_1 \cdots dy_m} \int_{D(y)} f_X(\Xi_x(x,t)) \exp\left[-\int_0^t \psi_X(\hat{x}(\tau))d\tau\right] dx_1 \cdots dx_n. \tag{8.55}$$

Then performing a change of variables on (8.55) and using (8.15) gives

$$\rho_Y(y,t) = \frac{d^m}{dy_1 \cdots dy_m} \int_{D(\Xi_y(y,t))} f_X(\Xi_x)\, d\Xi_{x_1} \cdots d\Xi_{x_n} \tag{8.56}$$

$$= \left|\frac{\partial y}{\partial \Xi_y}\right|^{-1} \frac{d^m}{d\Xi_{y_1} \cdots d\Xi_{y_m}} \int_{D(\Xi_y(y,t))} f_X(\Xi_x)\, d\Xi_{x_1} \cdots d\Xi_{x_n}. \tag{8.57}$$

The expression (8.57) matches that of $\hat{\rho}(x,t)$ in (8.52). Thus for the general case, the solution to the Liouville equation commutes with the Φ_{pdf} mapping.

8.5 The Liouville Equation for Control Systems

In the rest of this chapter, the ideas of the previous sections are extended to show that for Φ-related control systems, the abstracted system can capture the evolution of the uncertainty in the original system by an appropriate choice of control input. The relationship for Φ-related control systems with deterministic initial conditions is similar to that for the autonomous case, as shown in Figure 8.1. For stochastic initial conditions in Φ-related control system, the local relationship is shown in Figure 8.5.

First, the Liouville equation for control systems is described and its solution derived. Let S_X be a control system described locally by

$$\dot{x}(t) = F(x(t), u(t)) \tag{8.58}$$

where $x = (x_1, \ldots, x_n)$ and u is the control input. In this context, the time evolution of the pdf is a function of the control input. Given a point x and time t, the initial condition Ξ can be found only if the control input $u(\cdot)$ is known for all $[0,t]$. In such a case, the system can be considered an autonomous system and the backward dynamics of (8.58) can be solved to find Ξ. Thus,

$$\Xi = \Xi(x, t, u(\cdot)) \tag{8.59}$$

and the Liouville equation becomes

$$\frac{\partial \rho(x,t,u(\cdot))}{\partial t} + \frac{\partial [\rho(x,t,u(\cdot))F(x,u(t))]}{\partial x} = 0. \tag{8.60}$$

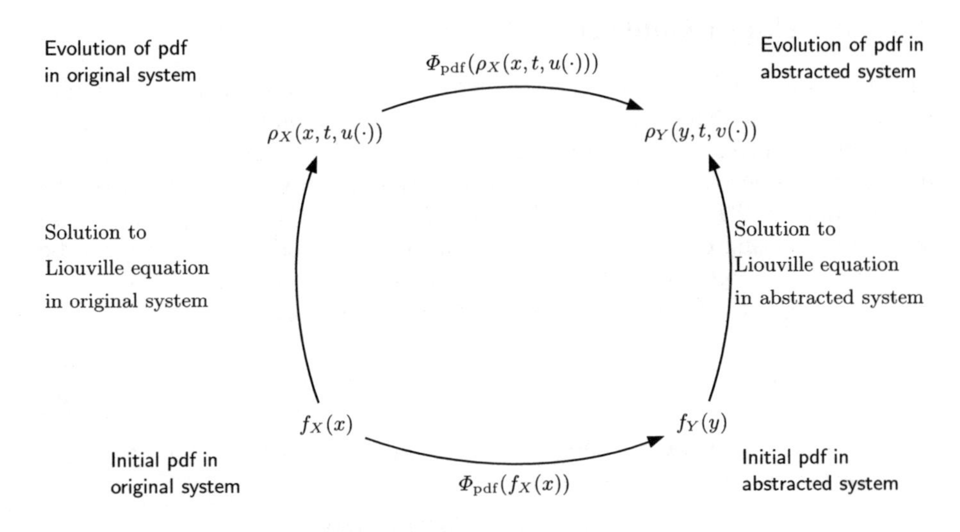

Fig. 8.5. The relationship between Φ-related control systems with uncertain initial conditions

The solution to the Liouville equation can be found by utilizing the method of characteristics as in [12] by letting

$$\frac{dx}{ds} = g(x, u(t)) \tag{8.61}$$

$$\frac{dt}{ds} = 1 \tag{8.62}$$

$$\frac{du}{ds} = 0. \tag{8.63}$$

The partial differential equation (8.60) becomes the ordinary differential equation

$$\frac{d\rho(x,t,u(\cdot))}{ds} = -\psi(x,u(t))\rho(x,t,u(\cdot)) \tag{8.64}$$

where

$$\psi(x,u(t)) = \sum_{i=1}^{n} \frac{\partial F_i(x,u(t))}{\partial x_i} \tag{8.65}$$

and $F_i(x,u(t))$ is the ith component of $F(x,u(t))$. The solution to (8.60) is

$$\rho(x,t,u(\cdot)) = f(\Xi(x,t,u(\cdot)))\exp\left[-\int_0^t \psi(\hat{x}(\tau),u(\tau))\,d\tau\right] \tag{8.66}$$

where $\hat{x}(\tau)$ denotes the trajectory starting at $\Xi(x,t,u(\cdot))$ at time zero and ending at x at time t. This solution is similar to that for the autonomous system, except that it shows explicitly the dependence of the pdf's evolution on the control input and that $u(\cdot)$ must be known for $[0,t]$.

8.6 Φ-Related Control Systems

In this section, the ideas of the previous sections are combined to show that the Φ_{pdf} mapping commutes, in some sense, with the solution to the Liouville equation for control systems. When an initial pdf for S_X is mapped through Φ_{pdf}, it then evolves according to the Liouville equation in a way that depends on the control input v. When the initial pdf evolves in S_X and then is mapped through Φ_{pdf}, the resulting expression can be obtained by a specific v and thus can be obtained by the solution to the Liouville equation in the abstracted system.

Suppose S_X and S_Y are Φ-related control systems described locally by

$$\dot{x} = F(x, u) \tag{8.67}$$

$$\dot{y} = G(y, v) \tag{8.68}$$

where $x = (x_1, \ldots, x_n) \in \mathbb{R}^n$ and $y = (y_1, \ldots, y_m) \in \mathbb{R}^m$. The divergences vector fields for each system are given by

$$\psi_X(x, u(t)) = \sum_{i=1}^{n} \frac{\partial F_i(x, u(t))}{\partial x_i} \tag{8.69}$$

$$\psi_Y(y, v(t)) = \sum_{i=1}^{m} \frac{\partial G_i(y, v(t))}{\partial y_i}. \tag{8.70}$$

Furthermore, let the initial condition $x(0)$ be uncertain with pdf $f_X(x)$ and let $u(\cdot)$ be known. Using the notation of Section 8.5, given points x and y at some time t, the initial conditions for each system are functions of their respective control inputs. That is,

$$\Xi_x = \Xi_x(x, t, u(\cdot)) \tag{8.71}$$

$$\Xi_y = \Xi_y(y, t, v(\cdot)) \tag{8.72}$$

where $u(\cdot)$ and $v(\cdot)$ must be known for $[0, t]$.

Given an initial pdf for the original system, $f_X(x)$, using (8.21) and mapping through Φ_{pdf} provides the corresponding pdf for the abstracted system

$$f_Y(y) = \frac{d^m}{dy_1 \cdots dy_m} \int_{D(y)} f_X(x) dx_1 \cdots dx_n. \tag{8.73}$$

This pdf propagates according to

$$\rho_Y(y, t, v(\cdot)) = f_Y(\Xi_y(y, t, v(\cdot))) \exp\left[-\int_0^t \psi_Y(\hat{y}(\tau), v(\tau))\, d\tau\right] \tag{8.74}$$

$$= \left[\frac{d^m}{d\Xi_{y_1} \cdots d\Xi_{y_m}} \int_{D(\Xi_y(y,t,v(\cdot)))} f_X(x) dx_1 \cdots dx_n\right]$$

$$\exp\left[-\int_0^t \psi_Y(\hat{y}(\tau), v(\tau))\, d\tau\right] \tag{8.75}$$

$$= \frac{d^m}{dy_1 \cdots dy_m} \int_{D(\Xi_y(y,t,v(\cdot)))} f_X(x) dx_1 \cdots dx_n. \tag{8.76}$$

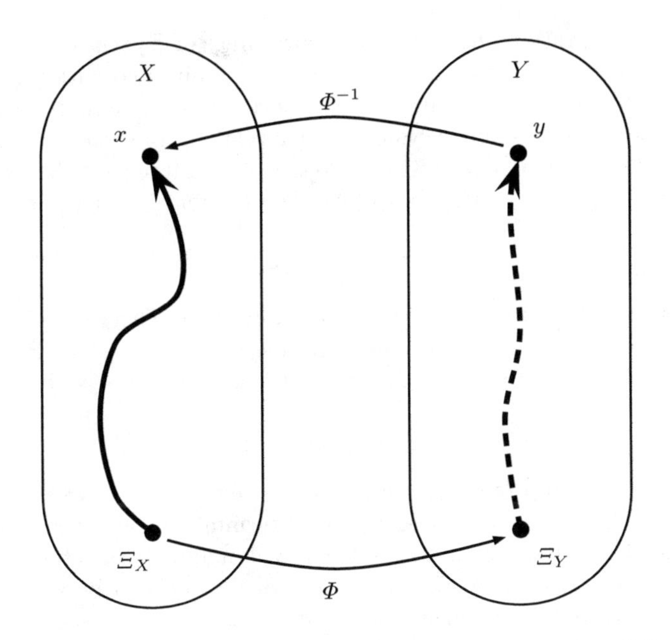

Fig. 8.6. The relationship between the endpoints of trajectories of S_X and S_Y

This evolution of the pdf depends on the control input $v(\cdot)$, which can be chosen freely.

On the other hand, the initial pdf $f_X(x)$ evolves as

$$\rho_X(x,t,u(\cdot)) = f_X(\Xi_x(x,t,u(\cdot))) \exp\left[-\int_0^t \psi_X(\hat{x}(\tau),u(\tau))\,d\tau\right]. \tag{8.77}$$

Mapping this evolution through Φ_{pdf} yields

$$\hat{\rho}_Y(y,t) = \frac{d^m}{dy_1\cdots dy_m}\int_{D(y)} \rho_X(x,t,u(\cdot))dx_1\cdots dx_n \tag{8.78}$$

$$= \frac{d^m}{dy_1\cdots dy_m}\int_{D(y)} f_X(\Xi_x(x,t,u(\cdot))) \exp\left[-\int_0^t \psi_X(\hat{x}(\tau),u(\tau))\,d\tau\right] dx_1\cdots dx_n. \tag{8.79}$$

Note that this expression (8.79) does not depend on v as an independent variable. Performing a change of variables on (8.79) gives

$$\hat{\rho}_Y(y,t) = \frac{d^m}{dy_1\cdots dy_m}\int_{D(\Xi_y)} f_X(\Xi_x)\,d\Xi_{x_1}\cdots d\Xi_{x_n} \tag{8.80}$$

where

$$D(\Xi_y) = \{x\in X : \Phi_1(x)\le \Xi_{y_1},\ldots,\Phi_m(x)\le \Xi_{y_m}\}. \tag{8.81}$$

The region of integration $D(\Xi_y)$ is taken over values of Ξ_y related to y, as shown in Figure 8.6. Given the endpoint y at time t, the initial condition Ξ_y does not depend on v, but is found through Φ. A trajectory exists connecting Ξ_y to y in Figure 8.6. Because the control systems are Φ-related, for each realization of the initial condition x_0 at time zero, there exists a control input for S_Y that depends on the control input $u(\cdot)$, such that $y(t) = \Phi(x(t))$ for all t. That is,

$$v = v(x_0, u(\cdot), \cdot). \tag{8.82}$$

The existence of this control input is guaranteed by Theorem 5.7. Thus, the control input v is not an independent variable as it is in (8.76). Rather, v depends explicitly on the initial condition $\Xi_x(x, t, u(\cdot))$ and the control input $u(\cdot)$. That is

$$v = v(\Xi_x(x, t, u(\cdot)), u(\cdot), \cdot). \tag{8.83}$$

By choosing the appropriate v in (8.76), the expression in (8.80) can be obtained. This relationship is similar to the deterministic case where the control input in the abstracted system can follow any Φ-mapped trajectory in the original system. The Φ_{pdf}-mapped "trajectory" of the pdf can be tracked using a feasible control input in the abstracted system.

9 Conclusion

This work has investigated abstraction of dynamical systems. Abstraction deals with the representation of a system using a simpler model which captures the important behavior of the original system. The motivating example throughout this work was the robotic car.

After a review of the mathematical preliminaries in Chapter 2, details of the robotic car modeling and one particular controller were reviewed in Chapter 3. Also in that chapter, results of a curvature estimator implemented for the author's Master's thesis were given. In Chapter 4, another controller was reviewed. This controller was designed for the unicycle rather than the car, and so must be converted in some way before being implemented on the car. This conversion was the motivation for studying abstraction and working on the problem: Can the controller designed for the unicycle be converted to one for the car?

The following chapters address this problem. In Chapter 5, abstraction concepts defined by previous authors were reviewed and shown to apply to the car/unicycle example. However, previous definitions didn't completely characterize the relationship between the car and unicycle, so traceability and ϵ-traceability were defined as new concepts to provide a more complete characterization. Chapters 6 and 7 applied these abstraction concepts to the control design problem.

The final chapter extended abstraction into the stochastic framework. If the initial condition for a system is uncertain, while the dynamics are still deterministic, how do the uncertainties propagate in the abstracted system? Does the concept of abstraction still apply to these systems? Intuitively, the results would seem to carry over to systems with uncertain initial conditions. Chapter 8 showed that abstraction results do indeed still apply.

The contributions of this work can be summarized as follows:

- The concept of abstraction is used to design a controller for a system based on an abstraction of its model. This concept is applied to the car/unicycle system and it shows that previous characterizations of abstraction do not fully capture certain behaviors that must be understood for control design.
- Abstraction of systems with uncertain initial conditions is studied in detail and it is shown that a commutative relationship exists between the systems. This shows that abstraction can be extended to the stochastic setting.

Abstraction of dynamical systems is a relatively new field of study and there are many relationships yet to be discovered. Two main areas that can be studied

P. Mellodge & P. Kachroo: Model Abstraction in Dynamical Systems, LNCIS 379, pp. 111–112, 2008.
springerlink.com

are the extension of abstraction to infinite dimensional systems and to stochastic systems.

For infinite dimensional systems, those systems modeled using partial differential equations (PDEs), can multiple PDEs be abstracted into one? Can a PDE be abstracted into an ordinary differential equation? It remains to be seen if the results studied here are applicable to such systems and whether the ideas must be modified to fit into the infinite dimensional framework.

For stochastic systems, can abstraction be extended to the case where the system dynamics themselves are uncertain? In such a situation, the Liouville equation becomes generalized to the Fokker-Plank equation and a diffusion term is introduced. Future work in this area includes studying system abstractions in which the dynamics and system parameters are stochastic.

Bibliography

[1] Abraham, R., Marsden, J., Ratiu, T.: Manifolds, Tensor Analysis and Applications. Springer, New York (1988)

[2] Aicardi, M., Casalino, G., Bicchi, A., Balestrino, A.: Closed loop steering of unicycle-like vehicles via Lyapunov techniques. IEEE Robotics and Automation Magazine, 27–35 (March 1995)

[3] Alexander, J.C., Maddocks, J.H.: On the kinematics of wheeled mobile robots. International Journal of Robotics Research 8(5), 15–27 (1989)

[4] Bay, J.S.: Fundamentals of Linear State Space Systems. WCB/McGraw-Hill, New York (1999)

[5] Bellaiche, A., Laumond, J.-P., Chyba, M.: Canonical nilpotent approximation of control systems: Application to nonholonomic motion planning. In: 32nd IEEE Conference on Decision and Control, San Antonio, TX (1993)

[6] Bellaiche, A., Laumond, J.-P., Jacobs, J.: Controllability of car-like robots and complexity of the motion planning problem. In: International Symposium on Intelligent Robotics, Bangalore, India, pp. 322–337 (1991)

[7] Brockett, R.W.: Lie algebras and lie groups in control theory. In: Geometric Methods in System Theory, pp. 43–82 (1973)

[8] Brockett, R.W.: Control theory and singular riemannian geometry. In: Hilton, P., Young, G. (eds.) New Directions in Applied Mathematics, pp. 11–27. Springer, New York (1981)

[9] Brockett, R.W.: Asymptotic stability and feedback stabilization. In: Brockett, R.W., Millman, R.S., Sussman, H.J. (eds.) Differential Geometric Control Theory, pp. 181–191. Birkhauser, Boston (1983)

[10] Canny, J.F.: The Complexity of Robot Motion Planning. The MIT Press, Cambridge (1988)

[11] Ehrendorfer, M.: The Liouville equation and its potential usefulness for the prediction of forecast skill. Part I: Theory. Monthly Weather Review 122 (April 1994)

[12] Ehrendorfer, M.: The Liouville equation and prediction of forecast skill. In: Grasman, J., Straten, G.v. (eds.) Predictability and Nonlinear Modeling in Natural Sciences and Economics, pp. 29–44. Kluwer, Dordrecht (1994)

[13] Fierro, R., Lewis, F.L.: Control of a nonholonomic mobile robot using neural networks. IEEE Transactions on Neural Networks 9(4), 589–600 (1998)

[14] Fliess, M., Levine, J., Martin, P., Rouchon, P.: On differentially flat nonlinear systems. In: Proceedings of the IFAC Nonlinear Control Systems Design Symposium, Bordeaux, France, pp. 408–412 (1992)

[15] Fliess, M., Levine, J., Martin, P., Rouchon, P.: Flatness and defect of nonlinear systems: Introductory theory and examples. International Journal of Control 61(6), 1327–1361 (1995)
[16] Fliess, M., Levine, J., Martin, P., Rouchon, P.: A Lie-Bäcklund approach to equivalence and flatness of nonlinear systems. IEEE Transactions on Automatic Control 44(5), 922–937 (1999)
[17] Friedberg, S.H., Insel, A.J., Spence, L.E.: Linear Algebra, 3rd edn. Prentice Hall, Upper Saddle River (1997)
[18] Ghanem, R., Masri, S., Pellissetti, M., Wolfe, R.: Identification and prediction of stochastic dynamical systems in a polynomial chaos basis. Computer Methods in Applied Mechanics and Engineering 194, 1641–1654 (2005)
[19] Ghosh, B.K., Loucks, E.P.: A perspective theory for motion and shape estimation in machine vision. SIAM Journal of Control and Optimization 33(5), 1530–1559 (1995)
[20] Heeger, D., Jepson, A.: Subspace methods for recovering rigid motion. International Journal of Computer Vision 7(2), 95–117 (1992)
[21] Isidori, A.: Nonlinear Control Systems, 3rd edn. Springer, New York (1995)
[22] Julius, A.A., Girard, A., Pappas, G.J.: Approximate bisimulation for a class of stochastic hybrid systems. In: Proceedings of the 2006 American Control Conference, Minneapolis, Minnesota, pp. 4724–4729 (June 2006)
[23] Kachroo, P.: Microprocessor-controlled small-scale vehicles for experiments in automated highway systems. The Korean Transport Policy Review 4(3), 145–178 (1997)
[24] Kachroo, P., Mellodge, P.: Mobile Robotic Car Design. McGraw-Hill, Inc., New York (2005)
[25] Kachroo, P., Tomizuka, M.: Vehicle control for automated highway systems for improved lateral maneuverability. In: IEEE International Conference on Systems, Man, and Cybernetics, Vancouver, B.C., Canada, vol. 1, pp. 777–782 (October 1995)
[26] Kirk, D.E.: Optimal Control Theory: An Introduction. Dover Publications, Inc., Mineola (2004)
[27] Kreyszig, E.: Advanced Engineering Mathematics, 7th edn. John Wiley & Sons, Inc., New York (1993)
[28] Lambert, A., Hamel, T., Le Fort-Piat, N.: A safe and robust path following planner for wheeled robots. In: Proc. Intl. Conf. on Intelligent Robots and Systems, Victoria, B.C., Canada, pp. 600–605 (October 1998)
[29] Laumond, J.-P.: Controllability of a multibody mobile robot. IEEE Transactions on Robotics and Automation (1993)
[30] Laumond, J.-P.: Robot Motion Planning and Control. Lecture Notes in Control and Information Sciences, vol. 229. Springer, New York (1998)
[31] Lee, T., Song, K., Lee, C., Teng, C.: Tracking control of unicycle-modeled mobile robots using a saturation feedback controller. IEEE Transactions on Control Systems Technology 9(2), 305–318 (2001)
[32] Li, Z., Canny, J.: Motion of two rigid bodies with rolling constraint. IEEE Transactions on Robotics and Automation (1990)
[33] Li, Z., Canny, J.F.: Nonholonomic Motion Planning. Kluwer Academic Publishers, Boston (1993)
[34] De Luca, A., Oriolo, G., Samson, C.: Feedback control of a nonholonomic car-like robot. In: Laumond, J.-P. (ed.) Robot Motion Planning and Control. Lecture Notes in Control and Information Sciences, vol. 229. Springer, New York (1998)

[35] Ma, Y., Kosecka, J., Sastry, S.S.: Vision guided navigation for a nonholonomic mobile robot. IEEE Transactions on Automatic Control 15(3), 521–536 (1999)
[36] Moret, E.N.: Dynamic modeling and control of a car-like robot. Master's thesis, Virginia Polytechnic Institute and State University, Bradley Department of Electrical and Computer Engineering (2003)
[37] Munkres, J.R.: Topology, 2nd edn. Prentice Hall, Upper Saddle River (2000)
[38] Murray, R.M., Sastry, S.S.: Steering nonholonomic systems in chained form. In: Proceedings of the 30th Conference on Decision and Control, Brighton, England, pp. 1121–1126 (December 1991)
[39] Murray, R.M., Sastry, S.S.: Nonholonomic motion planning: Steering using sinusoids. IEEE Transactions on Automatic Control 38(5), 700–716 (1993)
[40] Nakamura, Y., Savant, S.: Nonholonomic motion control of an autonomous underwater vehicle. In: IEEE International Workshop on Intelligent Robots and Systems, pp. 1254–1259 (November 1991)
[41] National Highway Traffic Safety Administration. The impact of driver inattention on near-crash/crash risk: An analysis using the 100-car naturalistic driving study data. Technical Report DOT HS 810 594, U.S. Department of Transportation, Washington, DC (April 2006)
[42] National Highway Traffic Safety Administration. Traffic safety facts 2006 data. Technical Report DOT HS 810 809, National Center for Statistics and Analysis, U.S. Department of Transportation, Washington, DC (March 2008)
[43] Neimark, Ju.I., Fufaev, N.A.: Dynamics of nonholonomic systems. In: Translations of Mathematical Monographs, vol. 33. American Mathematical Society, Providence (1972)
[44] Nijmeijer, H., van der Schaft, A.J.: Nonlinear Dynamical Control Systems. Springer, New York (1990)
[45] Oriolo, G., De Luca, A., Vendittelli, M.: WMR control via dynamic feedback linearization: Design, implementation, and experimental validation. IEEE Transactions on Control Systems Technology 10(6), 835–852 (2002)
[46] Pappas, G.J.: Bisimilar linear systems. Automatica 39, 2035–2047 (2003)
[47] Pappas, G.J., Lafferriere, G., Sastry, S.S.: Hierarchically consistent control systems. IEEE Transactions on Automatic Control 45(6), 1144–1160 (2000)
[48] Pappas, G.J., Simic, S.: Consistent abstractions of affine control systems. IEEE Transactions on Automatic Control 47(5), 745–756 (2002)
[49] Rouchon, P., Fliess, M., Levine, J., Martin, P.: Flatness, motion planning and trailer systems. In: Proceedings of the 32nd Conference on Decision and Control, San Antonio, TX, pp. 2700–2705 (December 1993)
[50] Rubinstein, R.Y.: Simulation and the Monte Carlo Method. John Wiley, New York (1981)
[51] Rudin, W.: Principles of Mathematical Analysis, 3rd edn. McGraw-Hill, Inc., New York (1976)
[52] Samson, C.: Control of chained systems application to path following and time varying point-stabilization of mobile robots. IEEE Transactions on Automatic Control 40(1), 64–77 (1995)
[53] Slotine, J.E., Li, W.: Applied Nonlinear Control. Prentice Hall, Englewood Cliffs (1991)
[54] de Sousa, C., Hemerly, E.M., Galvao, R.K.H.: Adaptive control for mobile robot using wavelet networks. IEEE Transactions on Systems, Man, and Cybernetics 32(4), 493–504 (2002)
[55] Sussmann, H.J.: Orbits of families of vector fields and integrability of distributions. Trans. Amer. Math. Soc. 180, 171–188 (1973)

[56] Sussmann, H.J.: Lie brackets, real analycity, and geometric control. In: Brockett, R.W., Millman, R., Sussmann, H.J. (eds.) Differential Geometric Control Theory. Birkhauser, Boston (1983)
[57] Sussmann, H.J., Jurdjevic, V.J.: Controllability of nonlinear systems. Journal of Differential Equations 12, 95–116 (1972)
[58] Tomasi, C., Kanade, T.: Shape and motion from image streams under orthography: A factorization method. International Journal of Computer Vision 9(2) (November 1992)
[59] van der Schaft, A.J.: Equivalence of dynamical systems by bisimulation. IEEE Transactions on Automatic Control 49(12) (December 2004)
[60] Vershik, A.M., Gershkovich, V.Ya.: Nonholonomic dynamical systems, geometry of distributions and variational problems. In: Arnol'd, V.I., Novikov, S.P. (eds.) Dynamical Systems VI. Encyclopedia of Mathematical Sciences, vol. 16. Springer, New York (1994)
[61] Xiu, D., Karniadakis, G.E.: The Wiener-Askey polynomial chaos for stochastic differential equations. SIAM Journal of Scientific Computing 24(2), 619–644 (2002)
[62] Yang, J., Kim, J.: Sliding mode control for trajectory tracking of nonholonomic wheeled mobile robots. IEEE Transactions on Robotics and Automation 15(3), 578–587 (1999)

Lecture Notes in Control and Information Sciences

Edited by M. Thoma, M. Morari

Further volumes of this series can be found on our homepage: springer.com

Vol. 379: Mellodge P.; Kachroo P.;
Model Abstraction in Dynamical Systems: Application to Mobile Robot Control
116 p. 2008 [978-3-540-70792-9]

Vol. 378: Femat R.; Solis-Perales G.;
Robust Synchronization of Chaotic Systems Via Feedback
xxx p. 2008 [978-3-540-69306-2]

Vol. 377: Patan K.
Artificial Neural Networks for the Modelling and Fault Diagnosis of Technical Processes
206 p. 2008 [978-3-540-79871-2]

Vol. 376: Hasegawa Y.
Approximate and Noisy Realization of Discrete-Time Dynamical Systems
245 p. 2008 [978-3-540-79433-2]

Vol. 375: Bartolini G.; Fridman L.; Pisano A.; Usai E. (Eds.)
Modern Sliding Mode Control Theory
465 p. 2008 [978-3-540-79015-0]

Vol. 374: Huang B.; Kadali R.
Dynamic Modeling, Predictive Control and Performance Monitoring
240 p. 2008 [978-1-84800-232-6]

Vol. 373: Wang Q.-G.; Ye Z.; Cai W.-J.; Hang C.-C.
PID Control for Multivariable Processes
264 p. 2008 [978-3-540-78481-4]

Vol. 372: Zhou J.; Wen C.
Adaptive Backstepping Control of Uncertain Systems
241 p. 2008 [978-3-540-77806-6]

Vol. 371: Blondel V.D.; Boyd S.P.; Kimura H. (Eds.)
Recent Advances in Learning and Control
279 p. 2008 [978-1-84800-154-1]

Vol. 370: Lee S.; Suh I.H.; Kim M.S. (Eds.)
Recent Progress in Robotics: Viable Robotic Service to Human
410 p. 2008 [978-3-540-76728-2]

Vol. 369: Hirsch M.J.; Pardalos P.M.; Murphey R.; Grundel D.
Advances in Cooperative Control and Optimization
423 p. 2007 [978-3-540-74354-5]

Vol. 368: Chee F.; Fernando T.
Closed-Loop Control of Blood Glucose
157 p. 2007 [978-3-540-74030-8]

Vol. 367: Turner M.C.; Bates D.G. (Eds.)
Mathematical Methods for Robust and Nonlinear Control
444 p. 2007 [978-1-84800-024-7]

Vol. 366: Bullo F.; Fujimoto K. (Eds.)
Lagrangian and Hamiltonian Methods for Nonlinear Control 2006
398 p. 2007 [978-3-540-73889-3]

Vol. 365: Bates D.; Hagström M. (Eds.)
Nonlinear Analysis and Synthesis Techniques for Aircraft Control
360 p. 2007 [978-3-540-73718-6]

Vol. 364: Chiuso A.; Ferrante A.; Pinzoni S. (Eds.)
Modeling, Estimation and Control
356 p. 2007 [978-3-540-73569-4]

Vol. 363: Besançon G. (Ed.)
Nonlinear Observers and Applications
224 p. 2007 [978-3-540-73502-1]

Vol. 362: Tarn T.-J.; Chen S.-B.; Zhou C. (Eds.)
Robotic Welding, Intelligence and Automation
562 p. 2007 [978-3-540-73373-7]

Vol. 361: Méndez-Acosta H.O.; Femat R.; González-Álvarez V. (Eds.):
Selected Topics in Dynamics and Control of Chemical and Biological Processes
320 p. 2007 [978-3-540-73187-0]

Vol. 360: Kozlowski K. (Ed.)
Robot Motion and Control 2007
452 p. 2007 [978-1-84628-973-6]

Vol. 359: Christophersen F.J.
Optimal Control of Constrained Piecewise Affine Systems
190 p. 2007 [978-3-540-72700-2]

Vol. 358: Findeisen R.; Allgöwer F.; Biegler L.T. (Eds.): Assessment and Future Directions of Nonlinear Model Predictive Control
642 p. 2007 [978-3-540-72698-2]

Vol. 357: Queinnec I.; Tarbouriech S.; Garcia G.; Niculescu S.-I. (Eds.):
Biology and Control Theory: Current Challenges
589 p. 2007 [978-3-540-71987-8]

Vol. 356: Karatkevich A.:
Dynamic Analysis of Petri Net-Based Discrete Systems
166 p. 2007 [978-3-540-71464-4]

Vol. 355: Zhang H.; Xie L.:
Control and Estimation of Systems with Input/Output Delays
213 p. 2007 [978-3-540-71118-6]

Vol. 354: Witczak M.:
Modelling and Estimation Strategies for Fault Diagnosis of Non-Linear Systems
215 p. 2007 [978-3-540-71114-8]

Vol. 353: Bonivento C.; Isidori A.; Marconi L.; Rossi C. (Eds.)
Advances in Control Theory and Applications
305 p. 2007 [978-3-540-70700-4]

Vol. 352: Chiasson, J.; Loiseau, J.J. (Eds.)
Applications of Time Delay Systems
358 p. 2007 [978-3-540-49555-0]

Vol. 351: Lin, C.; Wang, Q.-G.; Lee, T.H., He, Y.
LMI Approach to Analysis and Control of Takagi-Sugeno Fuzzy Systems with Time Delay
204 p. 2007 [978-3-540-49552-9]

Vol. 350: Bandyopadhyay, B.; Manjunath, T.C.; Umapathy, M.
Modeling, Control and Implementation of Smart Structures 250 p. 2007 [978-3-540-48393-9]

Vol. 349: Rogers, E.T.A.; Galkowski, K.; Owens, D.H.
Control Systems Theory and Applications for Linear Repetitive Processes
482 p. 2007 [978-3-540-42663-9]

Vol. 347: Assawinchaichote, W.; Nguang, K.S.; Shi P.
Fuzzy Control and Filter Design for Uncertain Fuzzy Systems
188 p. 2006 [978-3-540-37011-6]

Vol. 346: Tarbouriech, S.; Garcia, G.; Glattfelder, A.H. (Eds.)
Advanced Strategies in Control Systems with Input and Output Constraints
480 p. 2006 [978-3-540-37009-3]

Vol. 345: Huang, D.-S.; Li, K.; Irwin, G.W. (Eds.)
Intelligent Computing in Signal Processing and Pattern Recognition
1179 p. 2006 [978-3-540-37257-8]

Vol. 344: Huang, D.-S.; Li, K.; Irwin, G.W. (Eds.)
Intelligent Control and Automation
1121 p. 2006 [978-3-540-37255-4]

Vol. 341: Commault, C.; Marchand, N. (Eds.)
Positive Systems
448 p. 2006 [978-3-540-34771-2]

Vol. 340: Diehl, M.; Mombaur, K. (Eds.)
Fast Motions in Biomechanics and Robotics
500 p. 2006 [978-3-540-36118-3]

Vol. 339: Alamir, M.
Stabilization of Nonlinear Systems Using Receding-horizon Control Schemes
325 p. 2006 [978-1-84628-470-0]

Vol. 338: Tokarzewski, J.
Finite Zeros in Discrete Time Control Systems
325 p. 2006 [978-3-540-33464-4]

Vol. 337: Blom, H.; Lygeros, J. (Eds.)
Stochastic Hybrid Systems
395 p. 2006 [978-3-540-33466-8]

Vol. 336: Pettersen, K.Y.; Gravdahl, J.T.; Nijmeijer, H. (Eds.)
Group Coordination and Cooperative Control
310 p. 2006 [978-3-540-33468-2]

Vol. 335: Kozłowski, K. (Ed.)
Robot Motion and Control
424 p. 2006 [978-1-84628-404-5]

Vol. 334: Edwards, C.; Fossas Colet, E.; Fridman, L. (Eds.)
Advances in Variable Structure and Sliding Mode Control
504 p. 2006 [978-3-540-32800-1]

Vol. 333: Banavar, R.N.; Sankaranarayanan, V.
Switched Finite Time Control of a Class of Underactuated Systems
99 p. 2006 [978-3-540-32799-8]

Vol. 332: Xu, S.; Lam, J.
Robust Control and Filtering of Singular Systems
234 p. 2006 [978-3-540-32797-4]

Vol. 331: Antsaklis, P.J.; Tabuada, P. (Eds.)
Networked Embedded Sensing and Control
367 p. 2006 [978-3-540-32794-3]

Vol. 330: Koumoutsakos, P.; Mezic, I. (Eds.)
Control of Fluid Flow
200 p. 2006 [978-3-540-25140-8]

Vol. 329: Francis, B.A.; Smith, M.C.; Willems, J.C. (Eds.)
Control of Uncertain Systems: Modelling, Approximation, and Design
429 p. 2006 [978-3-540-31754-8]

Vol. 328: Loría, A.; Lamnabhi-Lagarrigue, F.; Panteley, E. (Eds.)
Advanced Topics in Control Systems Theory
305 p. 2006 [978-1-84628-313-0]

Vol. 327: Fournier, J.-D.; Grimm, J.; Leblond, J.; Partington, J.R. (Eds.)
Harmonic Analysis and Rational Approximation
301 p. 2006 [978-3-540-30922-2]

Vol. 326: Wang, H.-S.; Yung, C.-F.; Chang, F.-R.
H_∞ Control for Nonlinear Descriptor Systems
164 p. 2006 [978-1-84628-289-8]

Vol. 325: Amato, F.
Robust Control of Linear Systems Subject to Uncertain
Time-Varying Parameters
180 p. 2006 [978-3-540-23950-5]

Printing: Krips bv, Meppel, The Netherlands
Binding: Stürtz, Würzburg, Germany